STATISTICS FOR PSYCHOLOGISTS

An Intermediate Course

Brian S. Everitt
Institute of Psychiatry, King's College
University of London

LAWRENCE ERLBAUM ASSOCIATES, PUBLISHERS
2001   Mahwah, New Jersey   London

To Connie and Vera

Lawrence Erlbaum Associates, Inc., Publishers
10 Industrial Avenue
Mahwah, NJ 07430

Cover design by Kathryn Houghtaling Lacey

Library of Congress Cataloging-in-Publication Data

Everitt, Brian.
    Statistics for psychologists : an intermediate course / Brian S. Everitt.
        p.   cm.
    Includes bibliographical references and index.
    ISBN 0-8058-3836-8 (alk. paper)
    1. Psychology—Statistical methods.   I. Title.

 BF39 .E933   2001
 519.5′02415—dc21

                                                              00-065400

Books published by Lawrence Erlbaum Associates are printed on
acid-free paper, and their bindings are chosen for strength and durability.

Printed in the United States of America
10   9   8   7   6   5   4   3   2   1

Contents

Preface

Psychologists (and even those training to be psychologists) need little persuading that a knowledge of statistics is important in both the design of psychological studies and in the analysis of data from such studies. As a result, almost all undergraduate psychology students are exposed to an introductory statistics course during their first year of study at college or university. The details of the topics covered in such a course will no doubt vary from place to place, but they will almost certainly include a discussion of descriptive statistics, correlation, simple regression, basic inference and significance tests, p values, and confidence intervals. In addition, in an introductory statistics course, some nonparametric tests may be described and the analysis of variance may begin to be explored. Laudable and necessary as such courses are, they represent only the first step in equipping students with enough knowledge of statistics to ensure that they have a reasonable chance of progressing to become proficient, creative, influential, and even, perhaps, useful psychologists. Consequently, in their second or third years (and possibly in later postgraduate courses), many psychology students will be encouraged to learn more about statistics and, equally important, how to apply statistical methods in a sensible fashion. It is to these students that this book is aimed, and it is hoped that the following features of the text will help it reach its target.

1. The central theme is that statistics is about solving problems; data relevant to these problems are collected and analyzed to provide useful answers. To this end, the book contains a large number of real data sets arising from real problems. Numerical examples of the type that involve the skiing activities of belly dancers and politicians are avoided as far as possible.

2. Mathematical details of methods are largely confined to the displays. For the mathematically challenged, the most difficult of these displays can be regarded as "black boxes" and conveniently ignored. Attempting to understand the "input" to and "output" from the black box will be more important than understanding the minutiae of what takes place within the box. In contrast, for those students with less anxiety about mathematics, study of the relevant mathematical material (which on occasion will include the use of vectors and matrices) will undoubtedly help in their appreciation of the corresponding technique.

3. Although many statistical methods require considerable amounts of arithmetic for their application, the burden of actually performing the necessary calculations has been almost entirely removed by the development and wide availability of powerful and relatively cheap personal computers and associated statistical software packages. It is assumed, therefore, that most students will be using such tools when undertaking their own analyses. Consequently, arithmetic details are noticeable largely by their absence, although where a little arithmetic *is* considered helpful in explaining a technique, then it is included, usually in a table or a display.

4. There are many challenging data sets both in the text and in the exercises provided at the end of each chapter. (Answers, or hints to providing answers, to many of the exercises are given in an appendix.)

5. A (it is hoped) useful glossary is provided to both remind students about the most important statistical topics they will have covered in their introductory course and to give further details of some of the techniques covered in this text.

6. Before the exercises in each chapter, a short section entitled Computer hints is included. These sections are intended to help point the way to how to undertake the analyses reported in the chapter by using a statistical package. They are *not* intended to provide detailed computer instructions for the analyses. The book concentrates largely on two packages, SPSS and S-PLUS. (Details are to be found at www.spss.com and www.mathsoft.com. A useful reference for the SPSS software is Morgan and Griego, 1998, and for S-PLUS, Krause and Olson, 2000.) The main reason for choosing the former should be obvious—the package is widely used by working psychologists and by psychology students and is an extremely useful tool for undertaking a variety of analyses. But why S-PLUS because, at present at least, the package is not a favorite of psychologists? Well, there are two reasons: the first is that the author is an enthusiastic (although hardly an expert) user of the software; the second is that he sees S-PLUS as the software of the future. Admittedly, S-PLUS requires more initial effort to learn, but the reward for such investment is the ability to undertake both standard and nonstandard analyses routinely, with the added benefit of superb graphics. This is beginning to sound like a commercial, so

perhaps I will stop here! (Incidentally, for those readers who do not have access to the S-PLUS package but would like to try it out, there is a free cousin, R, details of which are given on www.cran.us.r-project.org.)

It is hoped that the text will be useful in a number of different ways, including the following.

1. As the main part of a formal statistics course for advanced undergraduate and postgraduate psychology students and also for students in other areas of the behavioral sciences. For example, Chapters 2–7 could for the basis of a course on linear models, with, perhaps, Chapter 10 being used to show how such models can be generalized.
2. As a supplement to an existing course—Chapters 8 and 9, for example, could be used to make students more aware of the wider aspects of nonparametric methods and the analysis of categorical data.
3. For self-study.
4. For quick reference.

I would like to thank Maria Konsolaki for running some SPSS examples for me, and my secretary, Harriet Meteyard, for her usual superb typing and general helpful advice in preparing the book.

Brian S. Everitt
London, June 2000

1

Statistics in Psychology: Data, Models, and a Little History

1.1. INTRODUCTION

Psychology is a uniquely diverse discipline, ranging from biological aspects of behaviour on the one hand to social psychology on the other, and from basic research to the applied professions of clinical, counselling, educational, industrial, organizational and forensic psychology.

—Andrew M. Colman, *Companion Encyclopedia of Psychology*, 1994.

As Dr. Colman states, psychology is indeed a diverse and fascinating subject, and many students are attracted to it because of its exciting promise of delving into many aspects of human behavior, sensation, perception, cognition, emotion, and personality, to name but a few. It probably comes as a disagreeable shock to many such students that they are often called on, early in their studies, to learn about statistics, because the subject (and, sadly, its practitioners, statisticians), are mostly seen as anything but exciting, and the general opinion seems to be that both should be avoided as far as possible.

However, this "head in the sand" attitude toward statistics taken by many a psychology student is mistaken. Both statistics (and statisticians) can be very

1

exciting (!), and anyway, a knowledge and understanding of the subject is essential at each stage in the often long, and frequently far from smooth, road from planning an investigation to publishing its results in a prestigious psychology journal. Statistical principles will be needed to guide the design of a study and the collection of data. The initial examination of data and their description will involve a variety of informal statistical techniques. More formal methods of estimation and significance testing may also be needed in building a model for the data and assessing its fit.

Nevertheless, most psychologists are not, and have no desire to be, statisticians (I can't think why). Consequently, questions have to be asked about what and how much statistics the average psychologist needs to know. It is generally agreed that psychologists should have some knowledge of statistics, at least, and this is reflected in the almost universal exposure of psychology students to an introductory course in the subject, covering topics such as

- descriptive statistics—histograms, means, variances, and standard deviations;
- elementary probability;
- the normal distribution;
- inference; t tests, chi-squared tests, confidence intervals, and so on;
- correlation and regression; and
- simple analyses of variance.

Although such a course often provides essential grounding in statistics, it frequently gives the wrong impression of what is and what is not of greatest importance in tackling real life problems. For many psychology students (and their teachers), for example, a p value is still regarded as the holy grail and almost the *raison d' être* of statistics (and providing one, the chief role of statisticians). Despite the numerous caveats issued in the literature, many psychologists still seem determined to experience joy on finding a p value of .049 and despair on finding one of .051. Again, psychology students may, on their introductory course, learn how to perform a t test (they may, poor things, even be made to carry out the arithmetic themselves), but they may still be ill equipped to answer the question, How can I summarize and understand the main features of this set of data?

The aim of this text is essentially twofold. First it will introduce the reader to a variety of statistical techniques not usually encountered in an introductory course; second, and equally, if not more importantly, it will attempt to transform the knowledge gained in such a course into a more suitable form for dealing with the complexities of real data. Readers will, for example, be encouraged to replace the formal use of significance tests and the rigid interpretation of p values with an approach that regards such tests as giving informal guidance on possible evidence of an interesting effect. Readers may even be asked to abandon the ubiquitous signficance test altogether in favor of, for example, a graphical display that makes the structure in the data apparent without any formal analyses. By building on and reshaping the statistical knowledge gained in their first-level course, students will be better equipped (it is hoped) to overcome the criticisms of much current statistical practice implied in the following quotations from two British statisticians:

> Most real-life statistical problems have one or more nonstandard features. There are no routine statistical questions; only questionable statistical routines.
>
> —Sir David Cox.

> Many statistical pitfalls lie in wait for the unwary. Indeed statistics is perhaps more open to misuse than any other subject, particularly by the nonspecialist. The misleading average, the graph with "fiddled axes," the inappropriate p-value and the linear regression fitted to nonlinear data are just four examples of horror stories which are part of statistical folklore.
>
> —Christopher Chatfield.

1.2. STATISTICS DESCRIPTIVE, STATISTICS INFERENTIAL, AND STATISTICAL MODELS

This text will be primarily concerned with the following overlapping components of the statistical analysis of data.

1. The initial examination of the data, with the aim of making any interesting patterns in the data more visible.
2. The estimation of parameters of interest.
3. The testing of hypotheses about parameters.
4. Model formulation, building, and assessment.

Most investigations will involve little clear separation between each of these four components, but for now it will be helpful to try to indicate, in general terms, the unique parts of each.

1.2.1. The Initial Examination of Data

The initial examination of data is a valuable stage of most statistical investigations, not only for scrutinizing and summarizing data, but often also as an aid to later model formulation. The aim is to clarify the general structure of the data, obtain simple descriptive summaries, and perhaps get ideas for a more sophisticated analysis. At this stage, distributional assumptions might be examined (e.g., whether the data are normal), possible outliers identified (i.e., observations very different from the bulk of the data that may be the result of, for example, a recording error), relationships between variables examined, and so on. In some cases the results from this stage may contain such an obvious message that more detailed analysis becomes largely superfluous. Many of the methods used in this preliminary analysis of the data will be *graphical*, and it is some of these that are described in Chapter 2.

1.2.2. Estimation and Significance Testing

Although in some cases an initial examination of the data will be all that is necessary, most investigations will proceed to a more formal stage of analysis that involves the estimation of population values of interest and/or testing hypotheses about particular values for these parameters. It is at this point that the beloved significance test (in some form or other) enters the arena. Despite numerous attempts by statisticians to wean psychologists away from such tests (see, e.g., Gardner and Altman, 1986), the p value retains a powerful hold over the average psychology researcher and psychology student. There are a number of reasons why it should not.

First, p value is poorly understood. Although p values appear in almost every account of psychological research findings, there is evidence that the general degree of understanding of the true meaning of the term is very low. Oakes (1986), for example, put the following test to 70 academic psychologists:

> Suppose you have a treatment which you suspect may alter performance on a certain task. You compare the means of your control and experimental groups (say 20 subjects in each sample). Further suppose you use a simple independent means t-test and your result is t = 2.7, df = 18, P = 0.01. Please mark each of the statements below as true or false.

> * You have absolutely disproved the null hypothesis that there is no difference between the population means.
> * You have found the probability of the null hypothesis being true.
> * You have absolutely proved your experimental hypothesis.
> * You can deduce the probability of the experimental hypothesis being true.
> * You know, if you decided to reject the null hypothesis, the probability that you are making the wrong decision.
> * You have a reliable experiment in the sense that if, hypothetically, the experiment were repeated a great number of times, you would obtain a significant result on 99% of occasions.

The subjects were all university lecturers, research fellows, or postgraduate students. The results presented in Table 1.1 are illuminating.

Under a relative frequency view of probability, all six statements are in fact *false*. Only 3 out of the 70 subjects came to the conclusion. The correct interpretation of the probability associated with the observed *t* value is

> the probability of obtaining the observed data (or data that represent a more extreme departure from the null hypothesis) if the null hypothesis is true.

TABLE 1.1

Frequencies and Percentages of "True" Responses in a Test of Knowledge
of p Values

Statement	f	%
1. The null hypothesis is absolutely disproved.	1	1.4
2. The probability of the null hypothesis has been found.	25	35.7
3. The experimental hypothesis is absolutely proved.	4	5.7
4. The probability of the experimental hypothesis can be deduced.	46	65.7
5. The probability that the decision taken is wrong is known.	60	85.7
6. A replication has a .99 probability of being significant.	42	60.0

Clearly the number of false statements described as true in this experiment would have been reduced if the true interpretation of a p value had been included with the six others. Nevertheless, the exercise is extremely interesting in highlighting the misguided appreciation of p values held by a group of research psychologists.

Second, a p value represents only limited information about the results from a study. Gardner and Altman (1986) make the point that the excessive use of p values in hypothesis testing, simply as a means of rejecting or accepting a particular hypothesis, at the expense of other ways of assessing results, has reached such a degree that levels of significance are often quoted alone in the main text and in abstracts of papers with no mention of other more relevant and important quantities. The implications of hypothesis testing—that there can always be a simple yes or no answer as the fundamental result from a psychological study—is clearly false, and used in this way hypothesis testing is of limited value.

The most common alternative to presenting results in terms of p values, in relation to a statistical null hypothesis, is to estimate the magnitude of some parameter of interest along with some interval that includes the population value of the parameter with some specified probability. Such *confidence intervals* can be found relatively simply for many quantities of interest (see Gardner and Altman, 1986 for details), and although the underlying logic of interval estimaton is essentially similar to that of significance tests, they do not carry with them the pseudoscientific hypothesis testing language of such tests. Instead they give a plausible range of values for the unknown parameter. As Oakes (1986) rightly comments,

the significance test relates to what the population parameter is *not*: the confidence interval gives a plausible range for what the parameter *is*.

So should the p value be abandoned completely? Many statistician would answer yes, but I think a more sensible response, at least for psychologists, would be a resounding "maybe." Such values should rarely be used in a purely confirmatory way, but in an exploratory fashion they can be useful in giving some informal guidance on the possible existence of an interesting effect, even when the required assumptions of whatever test is being used are known to be invalid. It is often possible to assess whether a p value is likely to be an underestimate or overestimate and whether the result is clear one way or the other. In this text, both p values and confidence intervals will be used; purely from a pragmatic point of view, the former are needed by psychology students because they remain of central importance in the bulk of the psychological literature.

1.2.3. The Role of Models in the Analysis of Data

Models imitate the properties of real objects in a simpler or more convenient form. A road map, for example, models part of the Earth's surface, attempting to reproduce the relative positions of towns, roads, and other features. Chemists use models of molecules to mimic their theoretical properties, which in turn can be used to predict the behavior of real objects. A good model follows as accurately as possible the relevant properties of the real object, while being convenient to use.

Statistical models allow inferences to be made about an object, or activity, or process, by modeling some associated observable data. Suppose, for example, a child has scored 20 points on a test of verbal ability, and after studying a dictionary for some time, scores 24 points on a similar test. If it is believed that studying the dictionary has caused an improvement, then a possible model of what is happening is

$$20 = \{\text{person's initial score}\}, \tag{1.1}$$

$$24 = \{\text{person's initial score}\} + \{\text{improvement}\}. \tag{1.2}$$

The improvement can be found by simply subtracting the first score from the second.

Such a model is, of course, very naive, because it assumes that verbal ability can be measured exactly. A more realistic representation of the two scores, which allows for a possible measurement error, is

$$x_1 = \gamma + \epsilon_1, \tag{1.3}$$

$$x_2 = \gamma + \delta + \epsilon_2, \tag{1.4}$$

where x_1 and x_2 represent the two verbal ability scores, γ represents the true initial measure of verbal ability, and δ represents the improvement score. The terms ϵ_1 and

ϵ_2 represent the measurement error. Here the improvement score can be *estimated* as $x_1 - x_2$.

A model gives a precise description of what the investigator assumes is occurring in a particular situation; in the case above it says that the improvement δ is considered to be independent of γ and is simply added to it. (An important point that should be noted here is that if you do not believe in a model, you should not perform operations and analyses on the data that assume it is true.)

Suppose now that it is believed that studying the dictionary does more good if a child already has a fair degree of verbal ability and that the various random influences that affect the test scores are also dependent on the true scores. Then an appropriate model would be

$$x_1 = \gamma \epsilon_1, \tag{1.5}$$

$$x_2 = \gamma \delta \epsilon_2. \tag{1.6}$$

Now the parameters are *multiplied* rather than added to give the observed scores, x_1 and x_2. Here δ might be estimated by dividing x_2 by x_1.

A further possibility is that there is a limit, λ, to improvement, and studying the dictionary improves performance on the verbal ability test by some proportion of the child's possible improvement, $\lambda - \gamma$. A suitable model would be

$$x_1 = \gamma + \epsilon_1, \tag{1.7}$$

$$x_2 = \gamma + (\lambda - \gamma)\delta + \epsilon_2. \tag{1.8}$$

With this model there is no way to estimate δ from the data unless a value of λ is given or assumed.

One of the principal uses of statistical models is to attempt to explain variation in measurements. This variation may be the result of a variety of factors, including variation from the measurement system, variation caused by environmental conditions that change over the course of a study, variation from individual to individual (or experimental unit to experimental unit), and so on. The decision about an appropriate model should be largely based on the investigator's prior knowledge of an area. In many situations, however, *additive, linear models*, such as those given in Eqs. (1.1) and (1.2), are invoked by default, because such models allow many powerful and informative statistical techniques to be applied to the data. *Analysis of variance techniques* (Chapters 3–5) and *regression analysis* (Chapter 6), for example, use such models, and in recent years *generalized linear models* (Chapter 10) have evolved that allow analogous models to be applied to a wide variety of data types.

Formulating an appropriate model can be a difficult problem. The general principles of model formulation are covered in detail in books on scientific method, but they include the need to collaborate with appropriate experts, to incorporate as much background theory as possible, and so on. Apart from those formulated entirely on *a priori* theoretical grounds, most models are, to some extent at least, based on an initial examination of the data, although completely empirical models are rare. The more usual intermediate case arises when a class of models is entertained *a priori*, but the initial data analysis is crucial in selecting a subset of models from the class. In a regression analysis, for example, the general approach is determined *a priori*, but a *scatter diagram* and a *histogram* (see Chapter 2) will be of crucial importance in indicating the "shape" of the relationship and in checking assumptions such as normality.

The formulation of a preliminary model from an initial examination of the data is the first step in the iterative, formulation–criticism, cycle of model building. This can produce some problems because formulating a model and testing it on the same data is not generally considered good science. It is always preferable to confirm whether a derived model is sensible by testing on new data. When data are difficult or expensive to obtain, however, then some model modification and assessment of fit on the original data is almost inevitable. Investigators need to be aware of the possible dangers of such a process.

The most important principal to have in mind when testing models on data is that of *parsimony*; that is, the "best" model is one that provides an adequate fit to data with the *fewest* number of parameters. This is often known as *Occam's razor*, which for those with a classical education is reproduced here in its original form:

Entia non stunt multiplicanda praeter necessitatem.

1.3.   TYPES OF STUDY

It is said that when Gertrude Stein lay dying, she roused briefly and asked her assembled friends, "Well, what's the answer?" They remained uncomfortably quiet, at which she sighed, "In that case, what's the question?"

Research in psychology, and in science in general, is about searching for the answers to particular questions of interest. Do politicians have higher IQs than university lecturers? Do men have faster reaction times than women? Should phobic patients be treated by psychotherapy or by a behavioral treatment such as flooding? Do children who are abused have more problems later in life than children who are not abused? Do children of divorced parents suffer more marital breakdowns themselves than children from more stable family backgrounds?

In more general terms, scientific research involves a sequence of asking and answering questions about the nature of relationships among variables (e.g., How does A affect B? Do A and B vary together? Is A significantly different from B?, and so on). Scientific research is carried out at many levels that differ in the types of question asked and, therefore, in the procedures used to answer them. Thus, the choice of which methods to use in research is largely determined by the kinds of questions that are asked.

Of the many types of investigation used in psychological research, the most common are perhaps the following:

- surveys;
- observational studies;
- quasi-experiments; and
- experiments.

Some *brief* comments about each of these four types are given below; a more detailed account is available in the papers by Stretch, Raulin and Graziano, and Dane, which all appear in the second volume of the excellent *Companion Encyclopedia of Psychology*; see Colman (1994).

1.3.1. Surveys

Survey methods are based on the simple discovery "that asking questions is a remarkably efficient way to obtain information from and about people" (Schuman and Kalton, 1985, p. 635). Surveys involve an exchange of information between researcher and respondent; the researcher identifies topics of interest, and the respondent provides knowledge or opinion about these topics. Depending upon the length and content of the survey as well as the facilities available, this exchange can be accomplished by means of written questionnaires, in-person interviews, or telephone conversations.

Surveys conducted by psychologists are usually designed to elicit information about the respondents' opinions, beliefs, attitudes and values. Some examples of data collected in surveys and their analysis are given in several later chapters.

1.3.2. Observational Studies

Many observational studies involve recording data on the members of naturally occurring groups, generally over a period of time, and comparing the rate at which a particular event of interest occurs in the different groups (such studies are often referred to as *prospective*). If, for example, an investigator was interested in the health effects of a natural disaster such as an earthquake, those who experienced the earthquake could be compared with a group of people who did not. Another commonly used type of observational study is the *case-control* investigation. Here a group of people (the *cases*) that all have a particular characteristic (a certain disease, perhaps) are compared with a group of people who do not have the characteristic (the *controls*), in terms of their past exposure to some event or *risk factor*. A recent example involved women that gave birth to very low birthweight infants (less than 1500 g), compared with women that had a child of normal birthweight with respect to their past caffeine consumption.

The types of analyses suitable for observational studies are often the same as those used for experimental studies (see Subsection 1.3.4). Unlike experiments, however, the lack of control over the groups to be compared in an observational study makes the interpretation of any difference between the groups detected in the study open to a variety of interpretations. In an investigation into the relationship between smoking and systolic blood pressure, for example, the researcher cannot allocate subjects to be smokers and nonsmokers (as would be required in an experimental approach), for obvious ethical reasons; instead, the systolic blood pressure of naturally occurring groups of individuals who smoke, and those who do not, are compared. In such a study any difference found between the blood pressure of the two groups would be open to three possible explanations.

1. Smoking *causes* a change in systolic blood pressure.
2. The level of blood pressure has a tendency to encourage or discourage smoking.
3. Some unidentified factors play a part in determining both the level of blood pressure and whether or not a person smokes.

1.3.3. Quasi-Experiments

Quasi-experimental designs resemble experiments proper (see next section), but they are weak on some of the characteristics. In particular (and like the observational study), the ability to manipulate the groups to be compared is not under the investigator's control. Unlike the observational study, however, the quasi-experiment involves the intervention of the investigator in the sense that he or she applies a variety of different "treatments" to naturally occurring groups. In an investigation of the effectiveness of three different methods of teaching mathematics to 15 year olds, for example, a method might be given to *all* the members of a particular class in a school. The three classes that receive the different teaching methods would be selected to be similar to each other on most relevant variables, and the methods would be assigned to classes on a chance basis.

1.3.4. Experiments

Experimental studies include most of the work psychologists carry out on animals and many of the investigations performed on human subjects in laboratories. The essential feature of an experiment is the large degree of control in the hands of the experimenters—in designed experiments, the experimenter deliberately changes the levels of the experimental factors to induce variation in the measured quantities, to lead to a better understanding of the relationship between experimental factors and the response variable. And, in particular, experimenters control the manner in which subjects are allocated to the different levels of the experimental factors. In a comparison of a new treatment with one used previously, for example, the researcher would have control over the scheme for allocating subjects to treatments. The manner in which this control is exercised is of vital importance if the results of the experiment are to be valid. If, for example, subjects who are first to volunteer are all allocated to the new treatment, then the two groups may differ in level of motivation and so subsequently in performance. Observed treatment differences would be confounded with differences produced by the allocation procedure.

The method most often used to overcome such problems is *random allocation* of subjects to treatments. Whether a subject receives the new or the old treatment is decided, for example, by the toss of a coin. The primary benefit that randomization has is the chance (and therefore impartial) assignment of extraneous influences among the groups to be compared, and it offers this control over such influences whether or not they are known by the experimenter to exist. Note that randomization does not claim to render the two samples equal with regard to these influences; if, however, the same procedure were applied in repeated samplings, equality would be achieved in the long run. This randomization ensures a lack of bias, whereas other methods of assignment may not.

In a properly conducted experiment (and this is the main advantage of such a study), the interpretation of an observed group difference is largely unambiguous; its *cause* is very likely to be the different treatments or conditions received by the two groups.

In the majority of cases, the statistical methods most useful for the analysis of data derived from experimental studies are the analysis of variance procedures described in Chapters 3–5.

1.4. TYPES OF DATA

The basic material that is the foundation of all psychological investigations is the measurements and observations made on a set of subjects. Clearly, not all measurement is the same. Measuring an individual's weight is qualitatively different from measuring his or her response to some treatment on a two-category scale, such as "improved" or "not improved," for example. Whatever measurements are made, they have to be objective, precise, and reproducible, for reasons nicely summarized in the following quotation from Fleiss (1986):

> The most elegant design of a study will not overcome the damage caused by unreliable imprecise measurement. The requirement that one's data be of high quality is at least as important a component of a proper study design as the requirement for randomization, double blinding, controlling where necessary for prognostic factors, and so on. Larger sample sizes than otherwise necessary, biased estimates and even biased samples are some of the untoward consequences of unreliable measurements that can be demonstrated.

Measurement scales are differentiated according to the degree of precision involved. If it is said that an individual has a high IQ, it is not as precise as the statement that the individual has an IQ of 151. The comment that a woman is tall is not as accurate as specifying that her height is 1.88 m. Certain characteristics of interest are more amenable to precise measurement than others. With the use of an accurate thermometer, a subject's temperature can be measured very precisely. Quantifying the level of anxiety or depression of a psychiatric patient, or assessing the degree of pain of a migraine sufferer is, however, far a more difficult task. Measurement scales may be classified into a hierarchy ranging from *categorical*, though *ordinal* to *interval*, and finally to *ratio scales*. Each of these will now be considered in more detail.

1.4.1. Nominal or Categorical Measurements

Nominal or categorical measurements allow patients to be classified with respect to some characteristic. Examples of such measurements are *marital status*, *sex*, and *blood group*.

The properties of a nominal scale are as follows.

1. Data categories are mutually exclusive (an individual can belong to only one category).
2. Data categories have no logical order—numbers may be assigned to categories but merely as convenient labels.

A nominal scale classifies without the categories being ordered. Techniques particularly suitable for analyzing this type of data are described in Chapters 9 and 10.

1.4.2. Ordinal Scales

The next level in the measurement hierarchy is the ordinal scale. This has one additional property over those of a nominal scale—a logical ordering of the categories. With such measurements, the numbers assigned to the categories indicate the *amount* of a characteristic possessed. A psychiatrist may, for example, grade patients on an anxiety scale as not anxious, mildly anxious, moderately anxious, or severely anxious, and he or she may use the numbers 0, 1, 2, and 3 to label the categories, with lower numbers' indicating less anxiety. The psychiatrist cannot infer, however, that the difference in anxiety between patients with scores of say 0 and 1 is the same as that between patients assigned scores 2 and 3. The scores on an ordinal scale do, however, allow patients to be *ranked* with respect to the characteristic being assessed. Frequently, however, measurements on an ordinal scale are described in terms of their mean and standard deviation. This is not appropriate if the steps on the scale are not known to be of equal length. Andersen (1990), Chapter 15, gives a nice illustration of why this is so.

The following are the properties of an ordinal scale.

1. Data categories are mutually exclusive.
2. Data categories have some logical order.
3. Data categories are scaled according to the amount of a particular characteristic they possess.

Chapter 8 covers methods of analysis particularly suitable for ordinal data.

1.4.3. Interval Scales

The third level in the measurement scale hierarchy is the interval scale. Such
scales possess all the properties of an ordinal scale, plus the additional property
that equal differences between category levels, on any part of the scale, reflect
equal differences in the characteristic being measured. An example of such a
scale is temperature on the Celsius or Fahrenheit scale; the difference between
temperatures of 80°F and 90°F is the same as between temperatures of 30°F and
40°F. An important point to make about interval scales is that the zero point is
simply another point on the scale; it does not represent the starting point of the
scale, nor the total absence of the characteristic being measured. The properties of
an interval scale are as follows.

1. Data categories are mutually exclusive.
2. Data categories have a logical order.
3. Data categories are scaled according to the amount of the characteristic they
   possess.
4. Equal differences in the characteristic are represented by equal differences
   in the numbers assigned to the categories.
5. The zero point is completely arbitrary.

1.4.4. Ratio Scales

The highest level in the hierarchy of measurement scales is the ratio scale. This
type of scale has one property in addition to those listed for interval scales, namely
the possession of a true zero point that represents the absence of the characteristic
being measured. Consequently, statements can be made both about differences on
the scale *and* the ratio of points on the scale. An example is weight, where not only
is the difference between 100 kg and 50 kg the same as between 75 kg and 25 kg,
but an object weighing 100 kg can be said to be twice as heavy as one weighing
50 kg. This is not true of, say, temperature on the Celsius or Fahrenheit scales,
where a reading of 100° does not represent twice the warmth of a temperature
of 50°. If, however, the temperatures were measured on the *Kelvin scale*, which
does have a true zero point, the statement about the ratio could be made.

The properties of a ratio scale are as follows.

1. Data categories are mutually exclusive.
2. Data categories have a logical order.
3. Data categories are scaled according to the amount of the characteristic they possess.
4. Equal differences in the characteristic are represented by equal differences in the numbers assigned to the categories.
5. The zero point represents an absence of the characteristic being measured.

An awareness of the different types of measurement that may be encountered in psychological studies is important, because the appropriate method of statistical analysis will often depend on the type of variable involved; this point is taken up again in later chapters.

A further classification of variable types is into *response* or *dependent* variables (also often referred to as *outcome* variables), and *independent* or *explanatory* variables (also occasionally called *predictor* variables). Essentially, the former are the variables measured by the investigator that appear on the left-hand side of the equation defining the proposed model for the data; the latter are variables thought to possibly affect the response variable and appear on the right-hand side of the model. It is the relationship between the dependent variable and the so-called independent variables (independent variables are often related; see comments and examples in later chapters) with which most studies in psychology are concerned. One further point; in some contexts, particularly that of analysis of variance (see Chapters 3–5), the independent variables are also often known as *factor variables*, or simply *factors*.

1.5. A LITTLE HISTORY

The relationship between psychology and statistics is a necessary one (honest!). A widely quoted remark by Galton is "that until the phenomena of any branch of knowledge have been submitted to measurement and number, it cannot assume the dignity of a science." And Galton was not alone in demanding measurement and numbers as a *sine qua non* for attaining the dignity of a science. Lord Kelvin is quoted as saying that one cannot understand a phenomenon until it is subjected to measurement, and Thorndike has said that whatever exists, exists in same amount, and could therefore eventually be subjected to measurement and counting.

Psychology has long striven to attain "the dignity of science" by submitting its observations to measurement and quantification. According to Singer (1979), David Hartley (1705–1757), in his major work, *Observations on Man* (1749), discussed the relevance of probability theory to the collection of scientific evidence, and argued for the use of mathematical and statistical ideas in the study of psychological processes.

A long-standing tradition in scientific psychology is the application of John Stuart Mill's experimental "method of difference" to the study of psychological problems. Groups of subjects are compared who differ with respect to the experimental treatment, but otherwise are the same in all respects. Any difference in outcome can therefore be attributed to the treatment. Control procedures such as randomization or matching on potentially confounding variables help bolster the assumption that the groups are the same in every way except the treatment conditions.

The experimental tradition in psychology has long been wedded to a particular statistical technique, namely the analysis of variance (ANOVA). The principles of experimental design and the analysis of variance were developed primarily by Fisher in the 1920s, but they took time to be fully appreciated by psychologists, who continued to analyze their experimental data with a mixture of graphical and simple statistical methods until well into the 1930s. According to Lovie (1979), the earliest paper that had ANOVA in its title was by Gaskill and Cox (1937). Other early uses of the technique are reported in Crutchfield (1938) and in Crutchfield and Tolman (1940).

Several of these early psychological papers, although paying lip service to the use of Fisher's analysis of variance techniques, relied heavily on more informal strategies of inference in interpreting experimental results. The year 1940, however, saw a dramatic increase in the use of analysis of variance in the psychological literature, and by 1943 the review paper of Garrett and Zubin was able to cite over 40 studies using an analysis of variance or covariance in psychology and education. Since then, of course, the analysis of variance in all its guises has become the main technique used in experimental psychology. An examination of 2 years of issues of the *British Journal of Psychology*, for example, showed that over 50% of the papers contain one or the other application of the analysis of variance.

Analysis of variance techniques are covered in Chapters 3, 4, and 5, and they are then shown to be equivalent to the multiple regression model introduced in Chapter 6.

1.6. WHY CAN'T A PSYCHOLOGIST BE MORE LIKE A STATISTICIAN (AND VICE VERSA)?

Over the course of the past 50 years, the psychologist has become a voracious consumer of statistical methods, but the relationship between psychologist and statistician is not always an easy, happy, or fruitful one. Statisticians complain that psychologists put undue faith in signficance tests, often use complex methods of analysis when the data merit only a relatively simple approach, and in general abuse many statistical techniques. Additionally, many statisticians feel that psychologists have become too easily seduced by user-friendly statistical software. These statisticians are upset (and perhaps even made to feel a little insecure) when their advice to plot a few graphs is ignored in favor of a multivariate analysis of covariance or similar statistical extravagence.

But if statisticians are at times horrified by the way in which psychologists apply statistical techniques, psychologists are no less horrified by many statisticians' apparent lack of awareness of what stresses psychological research can place on an investigator. A statistician may demand a balanced design with 30 subjects in each cell, so as to achieve some appropriate power for the analysis. But it is not the statistician who is faced with the frustration caused by a last-minute phone call from a subject who cannot take part in an experiment that has taken several hours to arrange. The statistician advising on a longitudinal study may call for more effort in carrying out follow-up interviews, so that no subjects are missed. It is, however, the psychologist who must continue to persuade people to talk about potentially distressing aspects of their lives, who must confront possibly dangerous respondents, or who arrives at a given (and often remote) address to conduct an interview, only to find that the person is not at home. In general, statisticians do not appear to appreciate the complex stories behind each data point in a psychological study. In addition, it is not unknown for statisticians to perform analyses that are statistically sound but psychologically naive or even misleading. An accomplished statistician, for example, once proposed an interpretation of findings regarding the benefits of nursery education, in which all subsequent positive effects could be accounted for in terms of the parents' choice of primary school. For once, it was psychologists who had to suppress a knowing smile; in the country for which the results were obtained, parents typically did not have any opportunity to choose the schools their children attended!

One way of examining the possible communication problems between psychologist and statistician is for each to know more about the language of the other. It is hoped this text will help in this process and enable young (and not so young) psychologists to learn more about the way statisticians approach the difficulties of data analysis, thus making their future consultations with statisticians more productive and less traumatic. (What is missing in this equation is, of course, a suitable *Psychology for Statisticians* text.)

1.7.   COMPUTERS AND STATISTICAL SOFTWARE

The development of computing and advances in statistical methodology have gone hand in hand since the early 1960s, and the increasing availability, power, and low cost of today's personal computer has further revolutionalized the way users of statistics work. It is probably hard for fresh-faced students of today, busily e-mailing everybody, exploring the delights of the internet, and in every other way displaying their computer literacy on the current generation of PCs, to imagine just what life was like for the statistician and the user of statistics in the days of simple electronic or mechanical calculators, or even earlier when large volumes of numerical tables were the only arithmetical aids available. It is a salutary exercise (of the "young people today hardly know they are born" variety) to illustrate with a little anecdote what things were like in the past.

Godfrey Thomson was an educational psychologist during the 1930s and 1940s. Professor A. E. Maxwell (personal communication) tells the story of Dr. Thomson's approach to the arithmetical problems faced when performing a factor analysis by hand. According to Maxwell, Godfrey Thomson and his wife would, early in the evening, place themselves on either side of their sitting room fire, Dr. Thomson equipped with several pencils and much paper, and Mrs. Thomson with a copy of *Barlow's Multiplication Tables*. For several hours the conversation would consist of little more than "What's 613.23 multiplied by 714.62?," and "438134.44": "What's 904.72 divided by 986.31?" and "0.91728," and so on.

Nowadays, increasingly sophisticated and comprehensive *statistics packages* are available that allow investigators easy access to an enormous variety of statistical techniques. This is not without considerable potential for performing very poor and often misleading analyses (a potential that many psychologists have grasped with apparent enthusiasm), but it would be foolish to underestimate the advantages of statistical software to users of statistics such as psychologists. In this text it will be assumed that readers will be carrying out their own analyses by using one or other of the many statistical packages now available, and we shall even end each chapter with limited computer hints for carrying out the reported analyses by using SPSS or S-PLUS. One major benefit of assuming that readers will undertake analyses by using a suitable package is that details of the arithmetic behind the methods will only rarely have to be given; consequently, descriptions of arithmetical calculations will be noticeable largely by their absence, although some tables *will* contain a little arithmetic where this is deemed helpful in making a particular point. Some tables will also contain a little mathematical nomenclature and, occasionally, even some equations, which in a number of places will use vectors and matrices. Readers who find this too upsetting to even contemplate should pass speedily by the offending material, regarding it merely as a black box and taking heart that understanding its input and output will, in most cases, be sufficient to undertake the corresponding analyses, and understand their results.

1.8. SAMPLE SIZE DETERMINATION

One of the most frequent questions asked when first planning a study in psychology is "how many subjects do I need?" Answering the question requires consideration of a number of factors, for example, the amount of time available, the difficulty of finding the type of subjects needed, and the cost of recruiting subjects. However, when the question is addressed statistically, a particular type of approach is generally used, in which these factors are largely ignored, at least in the first instance. Statistically, sample size determination involves

- identifying the response variable of most interest,
- specifying the appropriate statistical test to be used,
- setting a significance level,
- estimating the likely variability in the chosen response,
- choosing the power the researcher would like to achieve (for those readers who have forgotten, power is simply the probability of rejecting the null hypothesis when it is false), and
- committing to a magnitude of effect that the researcher would like to investigate. Typically the concept of a *psychologically relevant difference* is used, that is, the difference one would not like to miss declaring to be statistically significant.

These pieces of information then provide the basis of a statistical calculation of the required sample size. For example, for a study intending to compare the mean values of two treatments by means of a z test at the α significance level, and where the standard deviation of the response variable is know to be σ, the formula for the number of subjects required in each group is

$$n = \frac{2(Z_{\alpha/2} + Z_\beta)^2 \sigma^2}{\Delta^2}, \tag{1.9}$$

where Δ is the treatment difference deemed psychologically relevant, $\beta = 1 -$ power, and

- $Z_{\alpha/2}$ is the value of the normal distribution that cuts off an upper tail probability of $\alpha/2$; that is, if $\alpha = 0.05$, then $Z_{\alpha/2} = 1.96$; and
- Z_β is the value of the normal distribution that cuts off an upper tail probability of β; that is, if power $= 0.8$, so that $\beta = 0.2$, then $Z_\beta = 0.84$.

[The formula in Eq. (1.9) is appropriate for a two-sided test of size α.]

For example, for $\Delta = 1$, $\sigma = 1$, testing at the 5% level and requiring a power of 80%, the sample size needed in each group is

$$n = \frac{2 \times (1.96 + 0.84)^2 \times 1}{1} = 15.68. \tag{1.10}$$

So, essentially, 16 observations would be needed in each group.

Readers are encouraged to put in some other numerical values in Eq. (1.9) to get some feel of how the formula works, but its general characteristics are that n is an increasing function of σ and a decreasing function of Δ, both of which are intuitively understandable: if the variability of our response increases, then other things being equal, we ought to need more subjects to come to a reasonable conclusion. If we seek a bigger difference, we ought to be able to find it with fewer subjects.

In practice there are, of course, many different formulas for sample size determination, and nowadays software is widely available for sample size determination in many situations (e.g., nQuery Advisor, Statistical Solutions Ltd., www.statsolusa.com). However, one note of caution might be appropriate here, and it is that such calculations are often little more than "a guess masquerading as mathematics" (see Senn, 1997), as in practice there is often a tendency to "adjust" the difference sought and the power, in the light of what the researcher knows is a "practical" sample size in respect to time, money, and so on. Thus reported sample size calculations should frequently be taken with a large pinch of salt.

1.9. SUMMARY

1. Statistical principles are central to most aspects of a psychological investigation.
2. Data and their associated statistical analyses form the evidential parts of psychological arguments.
3. Significance testing is far from the be all and end all of statistical analyses, but it does still matter because evidence that can be discounted as an artifact of sampling will not be particularly persuasive. However, p values should not be taken too seriously; confidence intervals are often more informative.
4. A good statistical analysis should highlight those aspects of the data that are relevant to the psychological arguments, do so clearly and fairly, and be resistant to criticisms.
5. Experiments lead to the clearest conclusions about causal relationships.
6. Variable type often determines the most appropriate method of analysis.

EXERCISES

1.1. A well-known professor of experimental psychology once told the author, "If psychologists carry out their experiments properly, they rarely need statistics or statisticians."

1. Guess the professor!
2. Comment on his or her remark.

1.2. The Pepsi-Cola Company carried out research to determine whether people tended to prefer Pepsi Cola to Coca Cola. Participants were asked to taste two glasses of cola and then state which they preferred. The two glasses were not labeled as Pepsi or Coke for obvious reasons. Instead the Coke glass was labeled Q and the Pepsi glass was labelled M. The results showed that "more than half choose Pepsi over Coke" (Huck and Sandler, 1979, p. 11). Are there any alternative explanations for the observed difference, other than the taste of the two drinks? Explain how you would carry out a study to assess any alternative explanation you think possible.

1.3. Suppose you develop a headache while working for hours at your computer (this is probably a purely hypothetical possibility, but use your imagination). You stop, go into another room, and take two aspirins. After about 15 minutes your headache has gone and you return to work. Can you infer a definite causal relationship between taking the aspirin and curing the headache?

1.4. Attribute the following quotations about statistics and/or statisticians.

1. To understand God's thoughts we must study statistics, for these are a measure of his purpose.

2. You cannot feed the hungry on statistics.

3. A single death is a tragedy; a million deaths is a statistic.

4. I am not a number—I am a free man.

5. Thou shall not sit with statisticians nor commit a Social Science.

6. Facts speak louder than statistics.

7. I am one of the unpraised, unrewarded millions without whom statistics would be a bankrupt science. It is we who are born, marry, and who die in constant ratios.

2

Graphical Methods of Displaying Data

2.1. INTRODUCTION

A good graph is quiet and lets the data tell their story clearly and completely.

—H. Wainer, 1997.

According to Chambers, Cleveland, Kleiner, and Tukey (1983), "there is no statistical tool that is as powerful as a well chosen graph," and although this may be a trifle exaggerated, there is considerable evidence that there are patterns in data, and relationships between variables, that are easier to identify and understand from graphical displays than from possible alternatives such as tables. For this reason researchers who collect data are constantly encouraged by their statistical colleagues to make both a preliminary graphical examination of their data and to use a variety of plots and diagrams to aid in the interpretation of the results from more formal analyses. The prime objective of this approach is to communicate both to ourselves and to others.

But just what is a graphical display? A concise description is given by Tufte (1983).

Data graphics visually display measured quantities by means of the combined use of points, lines, a coordinate system, numbers, symbols, words, shading and colour.

Tufte estimates that between 900 billion (9×10^{11}) and 2 trillion (2×10^{12}) images of statistical graphics are printed each year.

Some of the advantages of graphical methods have been listed by Schmid (1954).

1. In comparison with other types of presentation, well-designed charts are more effective in creating interest and in appealing to the attention of the reader.
2. Visual relationships as portrayed by charts and graphs are more easily grasped and more easily remembered.
3. The use of charts and graphs saves time, because the essential meaning of large measures of statistical data can be visualized at a glance (like Chambers and his colleagues, Schmid may perhaps be accused of being prone to a little exaggeration here).
4. Charts and graphs provide a comprehensive picture of a problem that makes for a more complete and better balanced understanding than could be derived from tabular or textual forms of presentation.
5. Charts and graphs can bring out hidden facts and relationships and can stimulate, as well as aid, analytical thinking and investigation.

Schmid's last point is reiterated by the late John Tukey in his observation that "the greatest value of a picture is when it forces us to notice what we never expected to see."

During the past two decades, a wide variety of new methods for displaying data have been developed with the aim of making this particular aspect of the examination of data as informative as possible. Graphical techniques have evolved that will provide an overview, hunt for special effects in data, indicate outliers, identify patterns, diagnose (and criticize) models, and generally search for novel and unexpected phenomena. Good graphics will tell us a convincing story about the data. Large numbers of graphs may be required and computers will generally be needed to draw them for the same reasons that they are used for numerical analysis, namely that they are fast and accurate.

This chapter largely concerns the graphical methods most relevant in the initial phase of data analysis. Graphical techniques useful for diagnosing models and interpreting results will be dealt with in later chapters.

2.2. POP CHARTS

Newspapers, television, and the media in general are very fond of two very simple graphical displays, namely the *pie chart* and the *bar chart*. Both can be illustrated by using the data shown in Table 2.1, which show the percentage of people convicted

TABLE 2.1
Crime Rates for Drinkers and Abstainers

Crime	Drinkers	Abstainers
Arson	6.6	6.4
Rape	11.7	9.2
Violence	20.6	16.3
Stealing	50.3	44.6
Fraud	10.8	23.5

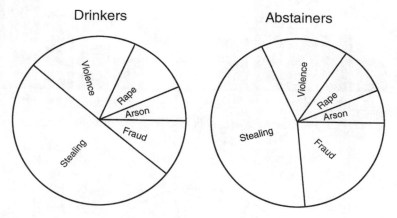

FIG. 2.1. Pie charts for drinker and abstainer crime percentages.

of five different types of crime. In the pie charts for drinkers and abstainers (see Figure 2.1), the sections of the circle have areas proportional to the observed percentages. In the corresponding bar charts (see Figure 2.2), percentages are represented by rectangles of appropriate size placed along a horizontal axis.

Despite their widespread popularity, both the general and scientific use of pie charts has been severely criticized. For example, Tufte (1983) comments that "tables are preferable to graphics for many small data sets. A table is nearly always better than a dumb pie chart; the only worse design than a pie chart is several of them . . . pie charts should never be used." A similar lack of affection is shown by Bertin (1981), who declares that "pie charts are completely useless," and more recently by Wainer (1997), who claims that "pie charts are the least useful of all graphical forms."

An alternative display that is always more useful than the pie chart (and often preferable to a bar chart) is the *dot plot*. To illustrate, we first use an example from Cleveland (1994). Figure 2.3 shows a pie chart of 10 percentages. Figure 2.4 shows

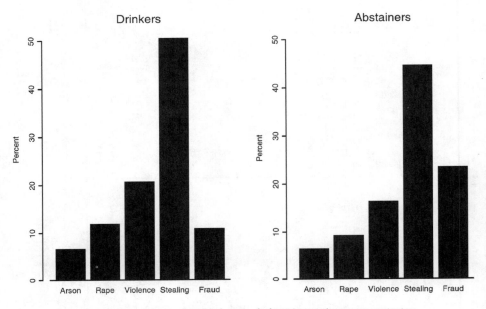

FIG. 2.2.   Bar charts for drinker and abstainer crime percentages.

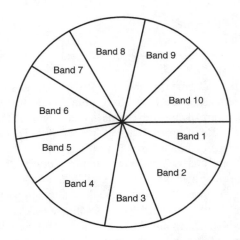

FIG. 2.3.   Pie chart for 10 percentages. (Reproduced with permission from Cleveland, 1994.)

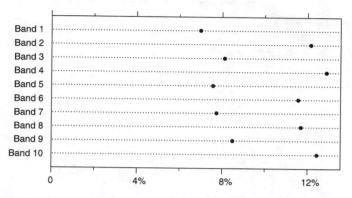

FIG. 2.4. Dot plot for 10 percentages. (Reproduced with permission from Cleveland, 1994.)

the alternative dot plot representation of the same values. Pattern perception is far more efficient for the dot plot than for the pie chart. In the former it is far easier to see a number of properties of the data that are either not apparent at all in the pie chart or only barely noticeable. First the percentages have a distribution with two modes (a *bimodal distribution*); odd-numbered bands lie around the value 8% and even-numbered bands around 12%. Furthermore, the shape of the pattern for the odd values as the band number increases is the same as the shape for the even values; each even value is shifted with respect to the preceding odd value by approximately 4%.

Dot plots for the crime data in Table 2.1 (see Figure 2.5) are also more informative than the pie charts in Figure 2.1, but a more exciting application of the dot plot is provided in Carr (1998). The diagram, shown in Figure 2.6, gives a particular contrast of brain mass and body mass for 62 species of animal. Labels and dots are grouped into small units of five to reduce the chances of matching error. In addition, the grouping encourages informative interpretation of the graph. Also note the thought-provoking title.

2.3. HISTOGRAMS, STEM-AND-LEAF PLOTS, AND BOX PLOTS

The data given in Table 2.2 shown the heights and ages of both couples in a sample of 100 married couples. Assessing general features of the data is difficult with the data tabulated in this way, and identifying any potentially interesting patterns is virtually impossible. A number of graphical displays of the data can help. For example, simple histograms of the heights of husbands, the heights of wives, and

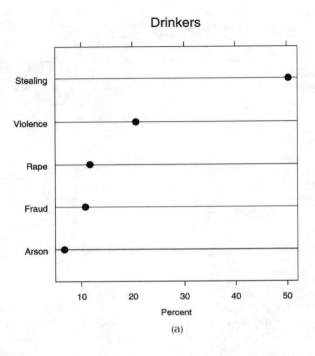

(a)

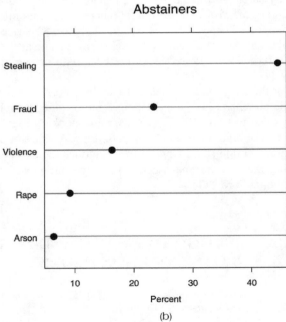

(b)

FIG. 2.5. Dot plots for (a) drinker and (b) abstainer crime percentages.

26

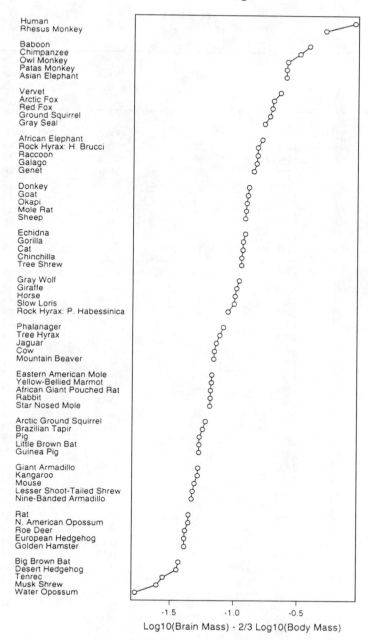

FIG. 2.6. Dot plot with positional linking (taken with permission from Carr, 1998).

TABLE 2.2
Heights and Ages of Married Couples

Husband's Age (years)	Husband's Height (mm)	Wife's Age (years)	Wife's Height (mm)	Husband's Age at First Marriage
49	1809	43	1590	25
25	1841	28	1560	19
40	1659	30	1620	38
52	1779	57	1540	26
58	1616	52	1420	30
32	1695	27	1660	23
43	1730	52	1610	33
47	1740	43	1580	26
31	1685	23	1610	26
26	1735	25	1590	23
40	1713	39	1610	23
35	1736	32	1700	31
35	1799	35	1680	19
35	1785	33	1680	24
47	1758	43	1630	24
38	1729	35	1570	27
33	1720	32	1720	28
32	1810	30	1740	22
38	1725	40	1600	31
29	1683	29	1600	25
59	1585	55	1550	23
26	1684	25	1540	18
50	1674	45	1640	25
49	1724	44	1640	27
42	1630	40	1630	28
33	1855	31	1560	22
27	1700	25	1580	21
57	1765	51	1570	32
34	1700	31	1590	28
28	1721	25	1650	23
37	1829	35	1670	22
56	1710	55	1600	44
27	1745	23	1610	25
36	1698	35	1610	22
31	1853	28	1670	20
57	1610	52	1510	25
55	1680	53	1520	21
47	1809	43	1620	25
64	1580	61	1530	21
31	1585	23	1570	28

(Continued)

TABLE 2.2
(Continued)

Husband's Age (years)	Husband's Height (mm)	Wife's Age (years)	Wife's Height (mm)	Husband's Age at First Marriage
35	1705	35	1580	25
36	1675	35	1590	22
40	1735	39	1670	23
30	1686	24	1630	27
32	1768	29	1510	21
20	1754	21	1660	19
45	1739	39	1610	25
59	1699	52	1440	27
43	1825	52	1570	25
29	1740	26	1670	24
47	1731	48	1730	21
43	1755	42	1590	20
54	1713	50	1600	23
61	1723	64	1490	26
27	1783	26	1660	20
27	1749	32	1580	24
32	1710	31	1500	31
54	1724	53	1640	20
37	1620	39	1650	21
55	1764	45	1620	29
36	1791	33	1550	30
32	1795	32	1640	25
57	1738	55	1560	24
51	1666	52	1570	24
50	1745	50	1550	22
32	1775	32	1600	20
54	1669	54	1660	20
34	1700	32	1640	22
45	1804	41	1670	27
64	1700	61	1560	24
55	1664	43	1760	31
27	1753	28	1640	23
55	1788	51	1600	26
41	1680	41	1550	22
44	1715	41	1570	24
22	1755	21	1590	21
30	1764	28	1650	29
53	1793	47	1690	31
42	1731	37	1580	23
31	1713	28	1590	28

(Continued)

TABLE 2.2
(Continued)

Husband's Age (years)	Husband's Height (mm)	Wife's Age (years)	Wife's Height (mm)	Husband's Age at First Marriage
36	1725	35	1510	26
56	1828	55	1600	30
46	1735	45	1660	22
34	1760	34	1700	23
55	1685	51	1530	34
44	1685	39	1490	27
45	1559	35	1580	34
48	1705	45	1500	28
44	1723	44	1600	41
59	1700	47	1570	39
64	1660	57	1620	32
34	1681	33	1410	22
37	1803	38	1560	23
54	1866	59	1590	49
49	1884	46	1710	25
63	1705	60	1580	27
48	1780	47	1690	22
64	1801	55	1610	37
33	1795	45	1660	17
52	1669	47	1610	23

the difference in husband and wife height may be a good way to begin to understand the data. The three histograms are shown in Figures 2.7 and 2.8. All the height distributions are seen to be roughly symmetrical and bell shaped, perhaps roughly normal? Husbands tend to be taller than their wives, a finding that simply reflects that men are on average taller than women, although there are a few couples in which the wife is taller; see the negative part of the x axis of the height difference histogram.

The histogram is generally used for two purposes: counting and displaying the distribution of a variable. According to Wilkinson (1992), however, "it is effective for neither." Histograms can often be misleading for displaying distributions because of their dependence on the number of classes chosen. Simple tallies of the observations are usually preferable for counting, particularly when shown in the form of a *stem-and-leaf plot*, as described in Display 2.1. Such a plot has the advantage of giving an impression of the shape of a variable's distribution, while retaining the values of the individual observations. Stem-and-leaf plots of the heights of husbands and wives and the height differences are shown in Figure 2.9.

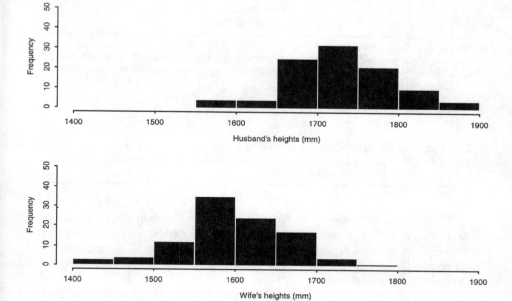

FIG. 2.7. Histograms of heights of husbands and their wives.

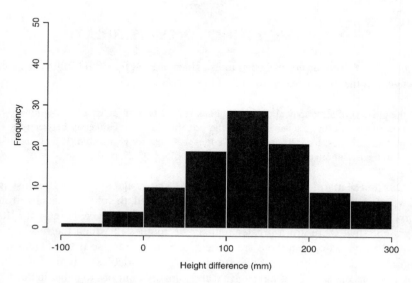

FIG. 2.8. Histogram of height difference for 100 married couples.

Display 2.1
Stem-and-Leaf Plots

To construct the simplest form of stem-and-leaf display of a set of observations, begin by choosing a suitable pair of adjacent digits in the heights data, the tens digit, and the units digit. Next, "split" each data value between the two digits. For example, the value 98 would be split as follows:

Data value   Split   stem   and   leaf
98   9/8   9   and   8

Then a separate line in the display is allocated for each possible string of leading digits (the *stems*). Finally, the trailing digit (the *leaf*) of each data value is written down on the line corresponding to its leading digit.

A further useful graphical display of a variable's distributional properties is the *box plot*. This is obtained from the *five-number summary* of a data set, the five numbers in question being the minimum, the lower quartile, the median, the upper quartile, and the maximum. The construction of a box plot is described in Display 2.2.

The box plots of the heights of husbands and their wives are shown in Figure 2.10. One unusually short man is identified in the plot, and three very short women. In addition, one rather tall woman is present among the wives.

2.4. THE SIMPLE SCATTERPLOT

The simple xy scatterplot has been in use since at least the 19th Century. Despite its age, it remains, according to Tufte (1983),

> the greatest of all graphical designs. It links at least two variables, encouraging and even imploring the viewer to assess the possible causal relationship between the plotted variables. It confronts causal theories that x causes y with empirical evidence as to the actual relationship between x and y.

Our first scatterplot, given in Figure 2.11(a), shows age of husband against age of wife for the data in Table 2.2. As might be expected, the plot indicates a strong correlation for the two ages. Adding the line $y = x$ to the plot, see Figure 2.11(b), highlights that there are a greater number of couples in which the husband is older than his wife, than there are those in which the reverse is true. Finally, in Figure 2.11(c) the bivariate scatter of the two age variables is framed with the observations on each. Plotting marginal and joint distributions together in this way is usually good data analysis practice; a further possibility for achieving this goal is shown in Figure 2.12.

```
N = 100    Median = 1727
Quartiles = 1690.5, 1771.5
```

Decimal point is 2 places to the right of the colon

```
    15 : 6888
    16 : 1223
    16 : 666777778888888899
    17 : 00000000000111111222222222233333334444444
    17 : 555556666677888899999
    18 : 00001112334
    18 : 5578
```

```
N = 100    Median = 1600
Quartiles = 1570, 1645
```

Decimal point is 2 places to the right of the colon

Low: 1410

```
    14 : 24
    14 : 99
    15 : 0011123344
    15 : 555566666777777788888889999999
    16 : 0000000011111111222233344444 4
    16 : 55566666667777778899
    17 : 001234
    17 : 6
```

```
N = 100    Median = 125
Quartiles = 79, 176.5
```

Decimal point is 2 places to the right of the colon

```
    -1 : 0
    -0 :
    -0 : 32
     0 : zz011333444
     0 : 566666777778888999
     1 : 00000111112222222222333333344444
     1 : 55566666677788999999
     2 : 0011233444
     2 : 5667889
```

FIG. 2.9. Stem-and-leaf plots of heights of husbands and wives and of height differences.

Display 2.2
Constructing a Box Plot

- The plot is based on five-number summary of a data set: 1, minimum; 2, lower quartile; 3, median; 4, upper quartile; 5, maximum.
- The distance between the upper and lower quartiles, the *interquartile range*, is a measure of the spread of a distribution that is quick to compute and, unlike the range, is not badly affected by outliers.
- The median, upper, and lower quartiles can be used to define rather arbitrary but still useful limits, L and U, to help identify possible outliers in the data:

$$U = UQ + 1.5 \times IQR,$$
$$L = LQ - 1.5 \times IQR.$$

Where UQ is the upper quartile, LQ is the lower quartile, and IQR the interquartile range, UQ–LQ.

- Observations outside the limits L and U are regarded as potential outliers and identified separately on the box plot (and known as *outside values*), which is constructed as follows:

To construct a box plot, a "box" with ends at the lower and upper quartiles is first drawn. A horizontal line (or some other feature) is used to indicate the position of the median in the box. Next, lines are drawn from each end of the box to the most remote observations that, however, are *not* outside observations as defined in the text. The resulting diagram schematically represents the body of the data *minus* the extreme observations. Finally, the outside observations are incorporated into the final diagram by representing them individually in some way (lines, stars, etc.)

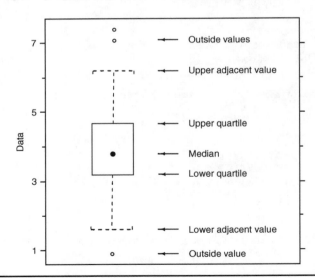

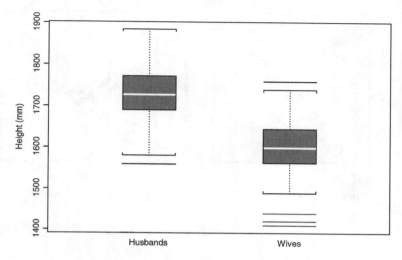

FIG. 2.10. Box plots of heights of husbands and wives.

The age difference in married couples might be investigated by plotting it against husband's age at marriage. The relevant scatterplot is shown in Figure 2.13. The points on the right-hand side of the line $x = 0$ in Figure 2.13 represent couples in which the husband is older; those to the left, couples in which a husband is younger than his wife. The diagram clearly illustrates the tendency of men marrying late in life to choose partners considerably younger than themselves. Furthermore, there are far fewer couples in which the wife is the older partner.

The relationship between the heights of the married couples might also be of interest. Figure 2.14(a) shows a plot of height of husband against height of wife. There is some indication of a positive association, but not one that is particularly strong. Again adding the line $y = x$ is informative; see Figure 2.14(b). There are few couples in which the wife is taller than her husband.

Finally, we might be interested in examining whether there is any evidence of an age difference–height difference relationship. The relevant scattergram is shown in Figure 2.15. The majority of marriages involve couples in which the husband is both taller and older than his wife. In only one of the 100 married couples is the wife both older and taller than her husband.

2.5. THE SCATTERPLOT MATRIX

In a set of data in which each observation involves more than two variables (*multivariate data*), viewing the scatterplots of each pair of variables is often a useful way to begin to examine the data. The number of scatterplots, however, quickly becomes daunting—for 10 variables, for example, there are 45 plots to consider.

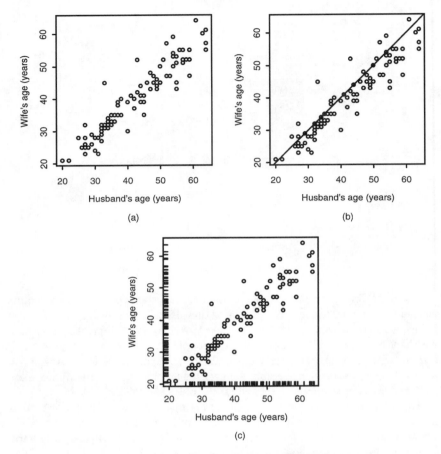

FIG. 2.11. Scatterplots of (a) ages of husbands and wives in 100 married couples; (b) with line $y = x$ added; (c) enhanced with observations on each variable.

Arranging the pairwise scatterplots in the form of a square grid, usually known as a *draughtsman's plot* or *scatterplot matrix*, can help in assessing all scatterplots at the same time.

Formally, a scatterplot matrix is defined as a square symmetric grid of bivariate scatterplots (Cleveland, 1994). This grid has p rows and p columns, each one corresponding to a different one of the p variables observed. Each of the grid's cells shows a scatterplot of two variables. Because the scatterplot matrix is symmetric about its diagonal, variable j is plotted against variable i in the ijth cell, and the same variables also appear in cell ji with the x and y axes of the scatterplot interchanged. The reason for including both the upper and lower triangles in the matrix, despite

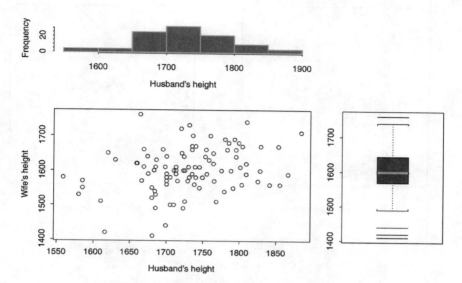

FIG. 2.12. Scatterplot of wife's height, against husband's height, showing marginal distributions of each variable.

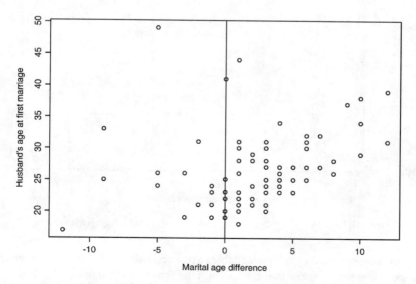

FIG. 2.13. Age difference of married couples plotted against husband's age at marriage.

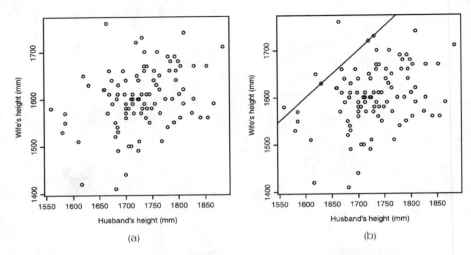

FIG. 2.14. Plots of (a) husband's height and wife's height in 100 married couples; (b) enhanced with the line $y = x$.

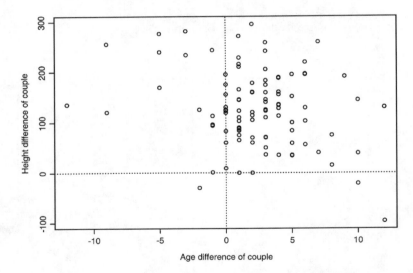

FIG. 2.15. Married couple's age difference and height difference.

the seeming redundancy, is that it enables a row and column to be visually scanned to see one variable against all others, with the scale for the one variable lined up along the horizontal or the vertical.

As our first illustration of a scatterplot matrix, Figure 2.16 shows such an arrangement for all the variables in the married couples' data given in Table 2.2.

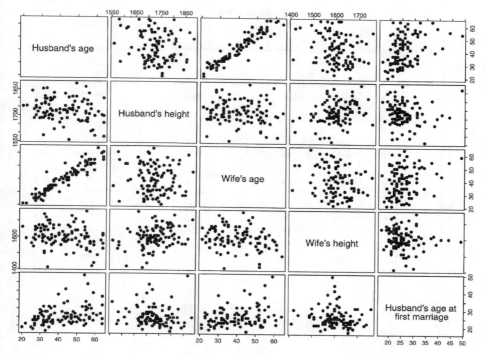

FIG. 2.16. Scatterplot matrix of heights and ages of married couples.

From this diagram the very strong relationship between age of husband and age of wife is apparent, as are the relative weaknesses of the associations between all other pairs of variables.

For a second example of the use of a scatterplot matrix, we shall use the data shown in Table 2.3. These data arise from an experiment in which 20 subjects had their response times measured when a light was flashed into each of their eyes through lenses of powers 6/6, 6/18, 6/36, and 6/60. (A lens of power a/b means that the eye will perceive as being at a feet an object that is actually positioned at b feet.) Measurements are in milliseconds, so the data are *multivariate*, but of a very special kind—the *same* variable is measured under different conditions. Such data are usually labeled *repeated measures* and will be the subject of detailed consideration in Chapters 5 and 7.

The scatterplot matrix of the data in Table 2.3 is shown in Figure 2.17. The diagram shows that measurements under particular pairs of lens strengths are quite strongly related, for example, 6/6 and 6/18, for both left and right eyes. In general, however, the associations are rather weak. The implications

TABLE 2.3
Visual Acuity and Lens Strength

Subject	Left Eye				Right Eye			
	L6/6	L6/18	L6/36	L6/60	R6/6	R6/18	R6/36	R6/60
1	116	119	116	124	12	117	114	122
2	110	110	114	115	106	112	110	110
3	117	118	120	120	120	120	120	124
4	112	116	115	113	115	116	116	119
5	113	114	114	118	114	117	116	112
6	119	115	94	116	100	99	94	97
7	110	110	105	118	105	105	115	115
8	117	119	119	123	110	110	115	111
9	116	120	123	124	115	115	120	114
10	118	118	119	120	112	120	120	119
11	120	119	118	121	110	119	118	122
12	110	112	115	118	120	123	123	124
13	118	120	122	124	120	118	116	122
14	118	120	120	122	112	120	122	124
15	110	110	110	120	120	119	118	124
16	112	112	114	118	110	110	108	120
17	112	112	120	122	105	110	115	105
18	105	105	110	110	118	120	118	120
19	110	120	1122	123	117	118	118	122
20	105	120	122	124	100	105	106	110

Note. Taken with permission from Crowder and Hand (1990).

of this pattern of associations for the analysis of these data will be taken up in Chapter 5.

2.6. ENHANCING SCATTERPLOTS

The basic scatterplot can accommodate only two variables, but there are ways in which it can be enhanced to display further variable values. The possibilities can be illustrated with a number of examples.

2.6.1. Ice Cream Sales

The data shown in Table 2.4 give the ice cream consumption over thirty 4-week periods, the price of ice cream in each period, and the mean temperature in each period. To display the values of all three variables on the same diagram, we use

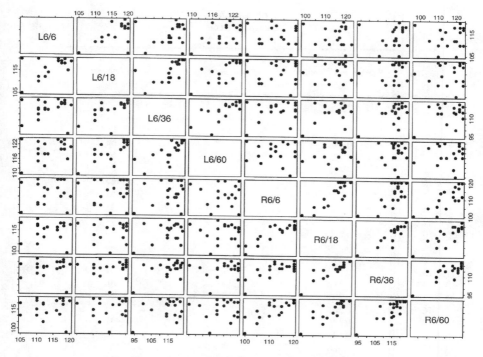

FIG. 2.17. Scatterplot matrix of visual acuity data.

what is generally known as a *bubble plot*. Here two variables are used to form a scatterplot in the usual way and then values of a third variable are represented by circles with radii proportional to the values and centered on the appropriate point in the scatterplot. The bubble plot for the ice cream data is shown in Figure 2.18. The diagram illustrates that price and temperature are largely unrelated and, of more interest, demonstrates that consumption remains largely constant as price varies and temperature varies below 50°F. Above this temperature, sales of ice cream increase with temperature but remain largely independent of price. The maximum consumption corresponds to the lowest price and highest temperature. One slightly odd observation is indicated in the plot; this corresponds to a month with low consumption, despite low price and moderate temperature.

2.6.2. Height, Weight, Sex, and Pulse Rate

As a further example of the enhancement of scatterplots, Figure 2.19 shows a plot of the height and weight of a sample of individuals with their gender indicated. Also included in the diagram are circles centered on each plotted point; the radii

TABLE 2.4
Ice Cream Data

Period	Y (Pints per Capita)	X_1 (Price per Pint)	X_2 (Mean Temperature, °F)
1	0.386	0.270	41
2	0.374	0.282	56
3	0.393	0.277	63
4	0.425	0.280	68
5	0.406	0.280	69
6	0.344	0.262	65
7	0.327	0.275	61
8	0.288	0.267	47
9	0.269	0.265	32
10	0.256	0.277	24
11	0.286	0.282	28
12	0.298	0.270	26
13	0.329	0.272	32
14	0.318	0.287	40
15	0.381	0.277	55
16	0.381	0.287	63
17	0.470	0.280	72
18	0.443	0.277	72
19	0.386	0.277	67
20	0.342	0.277	60
21	0.319	0.292	44
22	0.307	0.287	40
23	0.284	0.277	32
24	0.326	0.285	27
25	0.309	0.282	28
26	0.359	0.265	33
27	0.376	0.265	41
28	0.416	0.265	52
29	0.437	0.268	64
30	0.548	0.260	71

of these circles represent the value of the change in the pulse rate of the subject after running on the spot for 1 minute. Therefore, in essence, Figure 2.19 records the values of four variables for each individual: height, weight, changes in pulse rate, and gender. Information was also available on whether or not each person in the sample smoked, and this extra information is included in Figure 2.20, as described in the caption. Figure 2.20 is very useful for obtaining an overall picture of the data. The increase in pulse rate after exercise is clearly greater for men than for women, but it appears to be unrelated to whether or not an individual smokes.

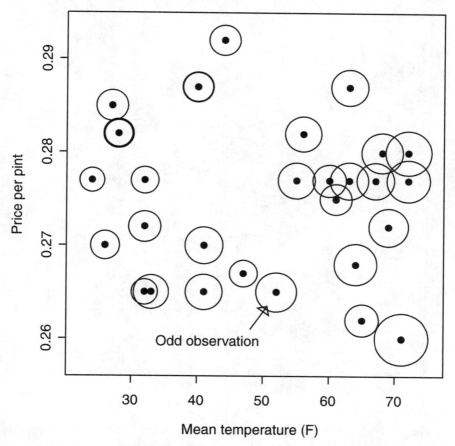

FIG. 2.18.   Bubble plot of ice cream data.

There is perhaps a relatively weak indication of a larger increase in pulse rate among heavier women, but no such relationship is apparent for men. Two men in the sample, one a smoker and one a nonsmoker, show a decrease in pulse rate after exercise.

2.6.3.   University Admission Rates

Bickel, Hammel, and O'Connell (1975) analyzed the relationship between admission rate and the proportion of women applying to the various academic departments at the University of California at Berkeley. The scatterplot of percentage of women applicants against percentage of applicants admitted is shown in Figure 2.21; boxes, the sizes of which indicate the relative number of applicants,

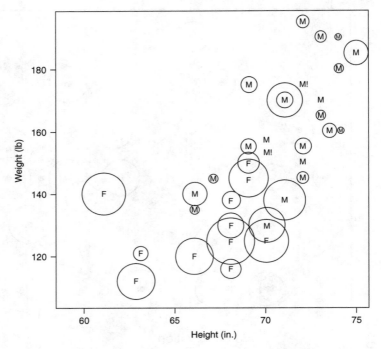

FIG. 2.19. Scatterplot of height and weight for male (M) and female (F) participants: the radii of the circles represent a change in pulse rate; ! indicates a decrease in pulse rate.

enhance the plot. The negative correlation indicated by the scatterplot is due almost exclusively to a trend for the large departments. If only a simple scatterplot had been used here, vital information about the relationship would have been lost.

2.7.   COPLOTS AND TRELLIS GRAPHICS

The *conditioning plot* or *coplot* is a particularly powerful visualisation tool for studying how two variables are related, *conditional* on one or more other variables being held constant. There are several varieties of conditioning plot; the differences are largely a matter of presentation rather than of real substance.

A simple coplot can be illustrated on the data from married couples given in Table 2.2. We shall plot the wife's height against the husband's height conditional on the husband's age. The resulting plot is shown in Figure 2.22. In this diagram

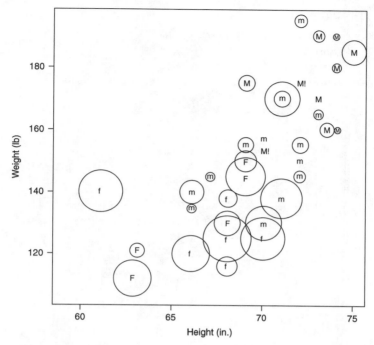

FIG. 2.20. Scatterplot of height and weight with additional information about sex, smoking status, and change in pulse rate: M, male smoker; F, female smoker; m, male nonsmoker; f, female nonsmoker; ! indicates a decrease in pulse rate.

the panel at the top of the figure is known as the *given panel*; the panels below are *dependence panels*. Each rectangle in the given panel specifies a range of values of the husband's age. On a corresponding dependence panel, the husband's height is plotted against the wife's height, for those couples in which the husband's age is in the appropriate interval. For age intervals to be matched to dependence panels, the latter are examined in order from left to right in the bottom row and again from left to right in subsequent rows. Here the coplot implies that the relationship between the two height measures is roughly the same for each of the six age intervals.

Coplots are a particular example of a more general form of graphical display known as *trellis graphics*, whereby any graphical display, not simply a scatterplot, can be shown for particular intervals of some variable. For example, Figure 2.23 shows a bubble plot of height of husband, height of wife, and age of wife (circles), conditional on husband's age for the data in Table 2.2. We leave interpretation of this plot to the reader!

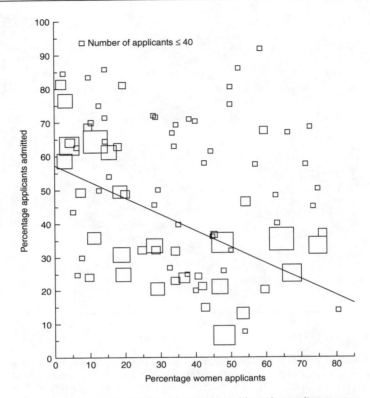

FIG. 2.21. Scatterplot of the percentage of female applicants versus percentage of applicants admitted for 85 departments at the University of California at Berkeley. (Reproduced with permission from Bickel et al., 1975.)

2.8. PROBABILITY PLOTTING

Many of the statistical methods to be described in later chapters are based on assuming either the normality of the raw data or the normality of some aspect of a fitted model. As readers will (it is hoped) recall from their introductory statistics course, normality implies that the observations in question arise from a bell-shaped probability distribution, defined explicitly by the formula

$$f(x) \; = \; \frac{1}{\sqrt{2\pi}\sigma} \; \exp \; \frac{-1}{2} \frac{(x - \mu)^2}{\sigma^2}, \qquad (2.1)$$

where μ is the mean of the distribution and σ^2 is its variance. Three examples of normal distributions are shown in Figure 2.24.

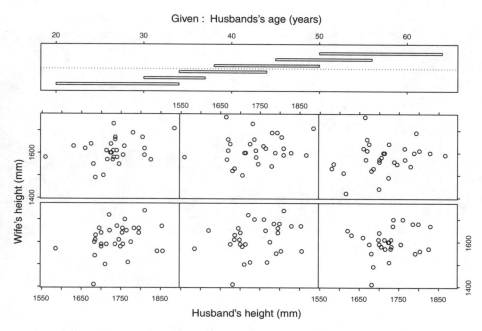

FIG. 2.22. Coplot of heights of husband and wife conditional on husband's age.

There are a variety of ways in which the normality assumption might be checked. Here we concentrate on the *normal probability plotting* approach, in which the observations are plotted against a set of suitably chosen percentiles of the *standard normal distribution*, that is, a normal distribution with $\mu = 0$ and $\sigma = 1$. Under the assumption of normality, this plot should approximate a straight line. Skewed populations are indicated by concave (left skew) and convex (right skew) plots. (For those readers who would like the full mathematical details behind such plots, see Display 2.3.) As a first illustration of this approach, Figure 2.25(a) shows a normal probability plot for a set of 100 observations generated from a normal distribution, and Figure 2.25(b) shows the corresponding plot for a set of 100 observations generated from an *exponential distribution*, a distribution with a high degree of skewness. In the first plot the points more or less lie on the required straight line; on the second, there is considerable departure from linearity.

Now let us look at some probability plots of real data, and here we shall use the data on married couples given in Table 2.2. Probability plots of all five variables are shown in Figure 2.26. Both height variables appear to have normal distributions, but all three age variables show some evidence of a departure from normality.

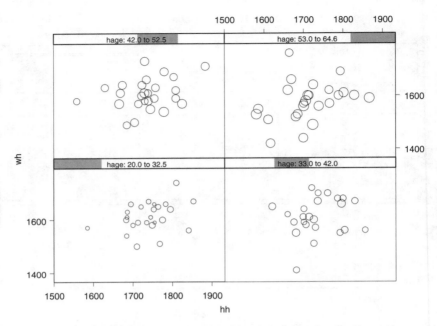

FIG. 2.23. Bubble plot of height of husband, height of wife, and age of wife (circles), conditional on age of husband.

2.9. GRAPHICAL DECEPTIONS AND GRAPHICAL DISASTERS

In general, graphical displays of the kind described in previous sections are extremely useful in the examination of data; indeed, they are almost essential both in the initial phase of data exploration and in the interpretation of the results from more formal statistical procedures, as will be seen in later chapters. Unfortunately, it is relatively easy to mislead the unwary with graphical material, and not all graphical displays are as honest as they should be! For example, consider the plot of the death rate per million from cancer of the breast, for several periods over the past three decades, shown in Figure 2.27. The rate appears to show a rather alarming increase. However, when the data are replotted with the vertical scale beginning at zero, as shown in Figure 2.28, the increase in the breast cancer death rate is altogether less startling. This example illustrates that undue exaggeration or compression of the scales is best avoided when one is drawing graphs (unless, of course, you are actually in the business of deceiving your audience).

A very common distortion introduced into the graphics most popular with newspapers, television, and the media in general is when *both* dimensions of a *two-dimensional figure* or *icon* are varied simultaneously in response to changes

Display 2.3
Normal Probability Plotting

- A normal probability plot involves plotting the n-ordered sample values, $y_{(1)}, \leq y_{(2)}, \ldots \leq y_{(n)}$, against the quantiles of a standard normal distribution, defined as

$$\Phi^{-1}\left[p_{(i)}\right],$$

where usually

$$p_i = \frac{i - 0.5}{n}, \qquad \Phi(x) = \int_{-\infty}^{x} \frac{1}{\sqrt{2\pi}} e^{-1/2 u^2}\, du.$$

- If the data arise from a normal distribution, the plot should be approximately linear.

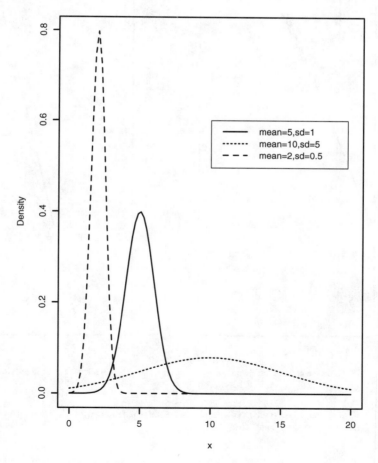

FIG. 2.24. Normal distributions.

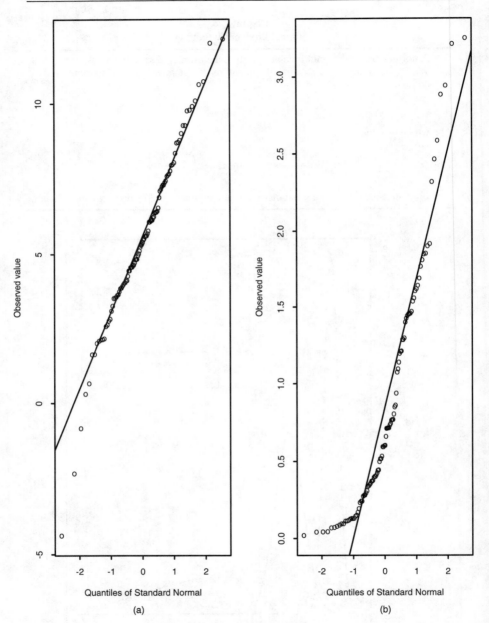

FIG. 2.25. Normal probability plots of (a) normal data, (b) data from an exponential distribution.

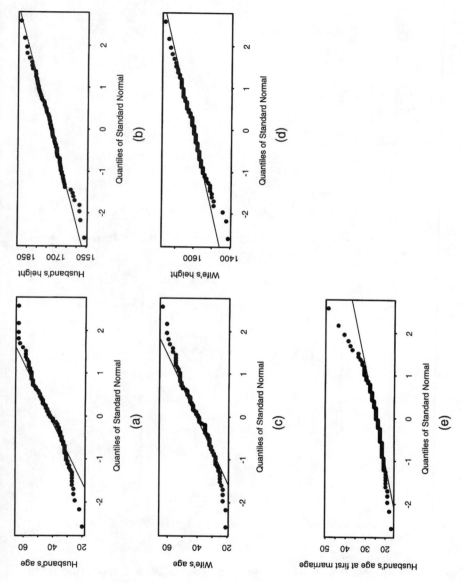

FIG. 2.26. Normal probability plots for data on married couples.

51

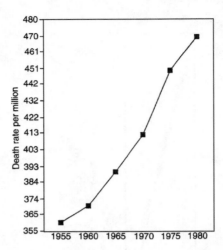

FIG. 2.27.   Death rates from cancer of the breast where the *y* axis does not include the origin.

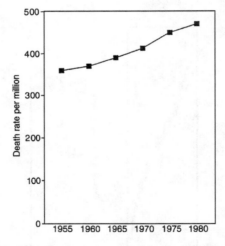

FIG. 2.28.   Death rates from cancer of the breast where the *y* axis does include the origin.

in a *single* variable. The examples shown in Figure 2.29, both taken from Tufte (1983), illustrate this point. Tufte quantifies the distortion with what he calls the *lie factor* of a graphical display, which is defined as

$$\text{lie factor} = \frac{\text{size of effect shown in graphic}}{\text{size of effect in data}}. \tag{2.2}$$

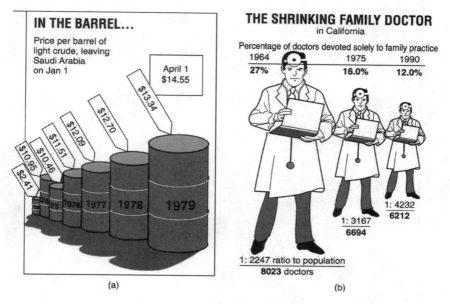

FIG. 2.29.   Graphics exhibiting lie factors of (a) 9.4 and (b) 2.8.

Lie factor values close to unity show that the graphic is probably representing the underlying numbers reasonably accurately. The lie factor for the oil barrels in Figure 2.29 is 9.4 because a 454% increase is depicted as 4280%. The lie factor for the shrinking doctors is 2.8.

A further example given by Cleveland (1994), and reproduced here in Figure 2.30, demonstrates that even the manner in which a simple scatterplot is drawn can lead to misperceptions about data. The example concerns the way in which judgment about the correlation of two variables made on the basis of looking at their scatterplot can be distorted by enlarging the area in which the points are plotted. The coefficient of correlation in the right-hand diagram in Figure 2.30 appears greater.

Some suggestions for avoiding graphical distortion taken from Tufte (1983) are as follows.

1. The representation of numbers, as physically measured on the surface of the graphic itself, should be directly proportional to the numerical quantities represented.
2. Clear, detailed, and thorough labeling should be used to defeat graphical distortion and ambiguity. Write out explanations of the data on the graphic itself. Label important events in the data.

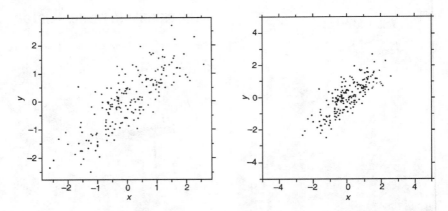

FIG. 2.30. Misjudgment of size of correlation caused by enlarging
the plotting area.

3. To be truthful and revealing, data graphics must bear on the heart of quantitative thinking: "compared to what?" Graphics must not quote data out of
   context.
4. Above all else, show the data.

Of course, not all poor graphics are deliberately meant to deceive. They are
just not as informative as they might be. An example is shown in Figure 2.31,
which originally appeared in Vetter (1980); its aim is to display the percentages of
degrees awarded to women in several disciplines of science and technology during
three time periods. At first glance the labels suggest that the graph is a standard
divided bar chart with the length of the bottom division of each bar showing the
percentage for doctorates, the length of the middle division showing the percentage
for master's degrees, and the top division showing the percentage for bachelor's
degrees. In fact, a little reflection shows that this is *not* correct, because it would
imply that in most cases the percentage of bachelor's degrees given to women is
generally lower than the percentage of doctorates. A closer examination of the
diagram reveals that the three values of the data for each discipline during each
time period are determined by the three adjacent vertical dotted lines. The top end
of the left-hand line indicates the value for doctorates, the top end of the middle
line indicates the value for master's degrees, and the top end of the right-hand line
indicates the position for bachelor's degrees.

Cleveland (1994) discusses other problems with the diagram in Figure 2.31 and
also points out that its manner of construction makes it hard to connect visually the
three values of a particular type of degree for a particular discipline, that is, to see
change through time. Figure 2.32 shows the same data as in Figure 2.31, replotted
by Cleveland in a bid to achieve greater clarity. It is now clear how the data are

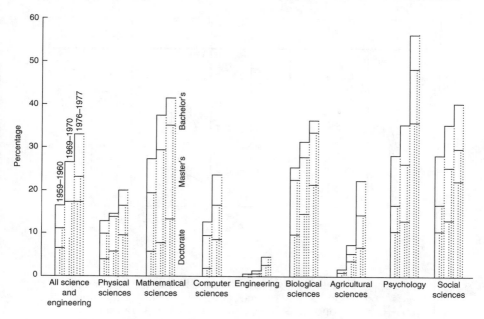

FIG. 2.31. Proportion of degrees in science and engineering earned by women in the periods 1959–1960, 1969–1970, and 1976–1977. (Reproduced with permission from Vetter, 1980.)

represented, and the design allows viewers to easily see the values corresponding to each degree in each discipline through time. Finally, the figure legend explains the graphs in a comprehensive and clear fashion. All in all, Cleveland appears to have produced a plot that would satisfy even that doyen of graphical presentation, Edward R. Tufte, in his demand that "excellence in statistical graphics consists of complex ideas communicated with clarity, precision and efficiency."

Being misled by graphical displays is usually a sobering but not a life-threatening experience. However, Cleveland (1994) gives an example in which using the wrong graph contributed to a major disaster in the American space program, namely the explosion of the *Challenger* space shuttle and the deaths of the seven people on board. To assess the suggestion that low temperature might affect the performance of the O-rings that sealed the joints of the rocket motor, engineers studied the graph of the data shown in Figure 2.33. Each data point was from a shuttle flight in which the O-rings had experienced thermal distress. The horizontal axis shows the O-ring temperature, and the vertical scale shows the number of O-rings that had experienced thermal distress. On the basis of these data, *Challenger* was allowed to take off when the temperature was 31°F, with tragic consequences.

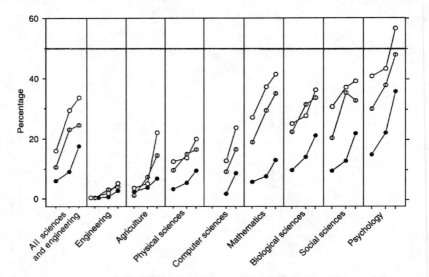

FIG. 2.32. Percentage of degrees earned by women for three degrees (bachelor's degree, master's degree, and doctorate), three time periods, and nine disciplines. The three points for each discipline and each degree indicate the periods 1959–1960, 1969–1970, and 1976–1977. (Reproduced with permission from Cleveland, 1994.)

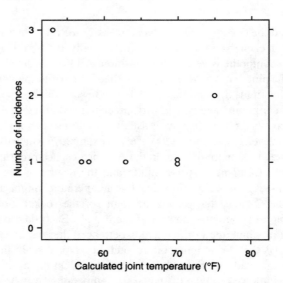

FIG. 2.33. Data plotted by space shuttle enginners the evening before the *Challenger* accident to determine the dependence of O-ring failure on temperature.

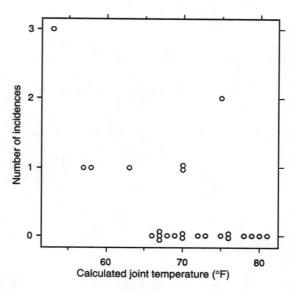

FIG. 2.34.   The complete set of O-ring data.

The data for no failures are not plotted in Figure 2.33. The engineers involved believed that these data were irrelevant to the issue of dependence. They
were mistaken, as shown by the plot in Figure 2.32, which includes *all* the
data. Here a pattern *does* emerge, and a dependence of failure on temperature is
revealed.

To end the chapter on a less sombre note, and to show that misperception
and miscommunication are certainly not confined to statistical graphics, see
Figure 2.35.

2.10.   SUMMARY

1. Graphical displays are an essential feature in the analysis of empirical data.
2. In some case a graphical "analysis" may be all that is required (or merited).
3. Stem-and-leaf plots are usually more informative than histograms for displaying frequency distributions.
4. Box plots display much more information about data sets and are very useful
   for comparing groups. In addition, they are useful for identifying possible
   outliers.
5. Scatterplots are the fundamental tool for examining relationships between
   variables. They can be enhanced in a variety of ways to provide extra information.

FIG. 2.35.   Misperception and miscommunication are sometimes
a way of life. (© The New Yorker collection 1961 Charles E. Martin
from cartoonbank.com. All Rights Reserved.)

6. Scatterplot matrices are a useful first step in examining data with more than
    two variables.
7. Beware graphical deception!

SOFTWARE HINTS

SPSS

Pie charts, bar charts, and the like are easily constructed from the Graph menu.
You enter the data you want to use in the chart, select the type of chart you
want from the Graph menu, define how the chart should appear, and then click
OK. For example, the first steps in producing a simple bar chart would be as
follows.

1. Enter the data you want to use to create the chart.
2. Click **Graph**, then click **Bar**. When you do this you will see the **Bar Charts**
    dialog box appear and the required chart can be constructed.
3. Once the initial chart has been created, it can be edited and refined very
    simply by double-clicking the chart and then making the required
    changes.

S-PLUS

In S-PLUS, there is a powerful graphical users interface whereby a huge variety of graphs can be plotted by using either the Graph menu or the two-dimensional (2D) or three-dimensional (3D) graphics palettes. For example, the necessary steps to produce a scatterplot are as follows.

1. Enter the data you want to use to create the scatterplot.
2. Click **Graph**; then click **2D Plot**.
3. Choose **Plot Type, ScatterPlot(x,y1,y2,...)**.
4. In the **Line/ScatterPlot** dialog box that now appears, choose the relevant data set and indicate the x and y variables for the plot.

Alternatively, when using the command line language, functions such as *plot, pairs, hist, pie, boxplot, dotplot* and many, many others can be used to produce graphical material. For example, if the vector *drink* contains the values of the percentage of crime rates for drinkers (see Table 2.1), the following command produces a pie chart:

pie(drink,density=-10,names=c("arson","Rape","Violence","Stealing","Fraud"))

[Detailed information about the *pie* function, or any of the other functions mentioned above, is available by using the help command; for example, *help(pie)*.]

Coplots and trellis graphics are currently only routinely available in S-PLUS. They can be implemented by use of the *coplot* function, or by using the 2D or 3D graphical palettes with the conditioning mode on. Details are given in Everitt and Rabe-Hesketh (2001).

EXERCISES

2.1. According to Cleveland (1994), "The histogram is a widely used graphical method that is at least a century old. But maturity and ubiquity do not guarantee the efficiency of a tool. The histogram is a poor method." Do you agree with Cleveland? Give your reasons.

2.2. Shortly after metric units of length were officially introduced in Australia, each of a group of 44 students was asked to guess, to the nearest meter, the width of the lecture hall in which they were sitting. Another group of 69 students in the same room were asked to guess the width in feet, to the nearest foot. (The true width of the hall was 13.1 m or 43.0 ft).

Guesses in meters:

8	9	10	10	10	10	10	10	11	11	11	11	12	12	13
13	13	14	14	14	15	15	15	15	15	15	15	15	16	16
16	17	17	17	17	18	18	20	22	25	27	35	38	40	

Guesses in feet:

24	25	27	30	30	30	30	30	30	32	32	33	34	34	34
35	35	36	36	36	37	37	40	40	40	40	40	40	40	40
40	41	41	42	42	42	42	43	43	44	44	44	45	45	45
45	45	45	46	46	47	48	48	50	50	50	51	54	54	54
55	55	60	60	63	70	75	80	94						

Construct suitable graphical displays for both sets of guesses to aid in throwing light on which set is more accurate.

2.3. Figure 2.36 shows the traffic deaths in a particular area before and after stricter enforcement of the speed limit by the police. Does the graph convince you that the efforts of the police have had the desired effect of reducing road traffic deaths? If not, why not?

2.4. Shown in Table 2.5 are the values of seven variables for 10 states in the USA.

TABLE 2.5

Information about 10 States in the USA

	Variable						
State	*1*	*2*	*3*	*4*	*5*	*6*	*7*
Alabama	3615	3624	2.1	69.05	15.1	41.3	20
California	21198	5114	1.1	71.71	10.3	62.6	20
Iowa	2861	4628	0.5	72.56	2.3	59.0	140
Mississippi	2341	3098	2.4	68.09	12.5	41.0	50
New Hampshire	812	4281	0.7	71.23	3.3	57.6	174
Ohio	10735	4561	0.8	70.82	7.4	53.2	124
Oregon	2284	4660	0.6	72.13	4.2	60.0	44
Pennsylvania	11860	4449	1.0	70.43	6.1	50.2	126
South Dakota	681	4167	0.5	72.08	1.7	52.3	172
Vermont	472	3907	0.6	71.64	5.5	57.1	168

Note. Variables are as follows: 1, population ($\times 1000$); 2, average per capita income ($); 3, illiteracy rate (% population); 4, life expectancy (years); 5, homicide rate (per 1000); 6, percentage of high school graduates; 7, average number of days per year below freezing.

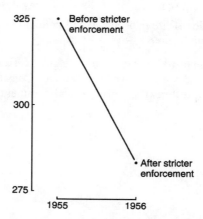

FIG. 2.36. Traffic deaths before and after the introduction of a stricter enforcement of the speed limit.

TABLE 2.6
Mortality Rates per 100,000 from Male Suicides

Country	Age Group				
	25–34	*35–44*	*45–54*	*55–64*	*65–74*
Canada	22	27	31	34	24
Israel	9	19	10	14	27
Japan	22	19	21	31	49
Austria	29	40	52	53	69
France	16	25	36	47	56
Germany	28	35	41	59	52
Hungary	48	65	84	81	107
Italy	7	8	11	18	27
Netherlands	8	11	18	20	28
Poland	26	29	36	32	28
Spain	4	7	10	16	22
Sweden	28	41	46	51	35
Switzerland	22	34	41	50	41
UK	10	13	15	17	22
USA	20	22	28	33	37

1. Construct a scatterplot matrix of the data, labeling the points by state name.

2. Construct a coplot of life expectancy and homicide rate conditional on the average per capita income.

2.5. Mortality rates per 100,000 from male suicides for a number of age groups and a number of countries are shown in Table 2.6. Construct side-by-side box plots for the data from different age groups, and comment on what the graphic says about the data.

3

Analysis of Variance I: The One-Way Design

3.1. INTRODUCTION

In a study of fecundity of fruit flies, per diem fecundity (number of eggs laid per female per day for the first 14 days of life) for 25 females of each of three genetic lines of the fruit fly *Drosophila melanogaster* was recorded. The results are shown in Table 3.1. The lines labeled RS and SS were selectively bred for resistance and for susceptibility to the pesticide, DDT, and the NS line as a nonselected control strain. Of interest here is whether the data give any evidence of a difference in fecundity of the three strains.

In this study, the effect of a single independent factor (genetic strain) on a response variable (per diem fecundity) is of interest. The data arise from what is generally known as a *one-way design*. The general question addressed in such a data set is, Do the populations giving rise to the different levels of the independent factor have different mean values? What statistical technique is appropriate for addressing this question?

3.2. STUDENT'S t TESTS

Most readers will recall Student's t tests from their introductory statistics course (those whose memories of statistical topics are just a little shaky can give

TABLE 3.1
Fecundity of Fruitflies

Resistant (RS)	Susceptible (SS)	Nonselected (NS)
12.8	38.4	35.4
21.6	32.9	27.4
14.8	48.5	19.3
23.1	20.9	41.8
34.6	11.6	20.3
19.7	22.3	37.6
22.6	30.2	36.9
29.6	33.4	37.3
16.4	26.7	28.2
20.3	39.0	23.4
29.3	12.8	33.7
14.9	14.6	29.2
27.3	12.2	41.7
22.4	23.1	22.6
27.5	29.4	40.4
20.3	16.0	34.4
38.7	20.1	30.4
26.4	23.3	14.9
23.7	22.9	51.8
26.1	22.5	33.8
29.5	15.1	37.9
38.6	31.0	29.5
44.4	16.9	42.4
23.2	16.1	36.6
23.6	10.8	47.4

themselves a reminder by looking at the appropriate definition in the glossary in Appendix A). The most commonly used form of this test addresses the question of whether the means of two distinct populations differ. Because interest about the fruit fly data in Table 3.1 also involves the question of differences in population means, is it possible that the straightforward application of the t test to each pair of strain means would provide the required answer to whether the strain fecundities differ? Sadly, this is an example of where putting two and two together arrives at the wrong answer. To explain why requires a little simple probability and algebra, and so the details are confined to Display 3.1. The results given there show that the consequences of applying a series of t tests to a one-way design with a moderate number of groups is very likely to be a claim of a significant difference, even when, in reality, no such difference exists. Even with only three groups as in the fruit fly example, the separate t tests approach increases the nominal 5%

Display 3.1
The Problem with Using Multiple t Tests

- The null hypothesis of interest is

$$H_0 : \mu_1 = \mu_2 = \cdots = \mu_k.$$

- Suppose the hypothesis is investigated by using a series of t tests, one for each pair of means.
- The total number of t tests needed is $N = k(k-1)/2$.
- Suppose that each t test is performed at significance level α, so that for each of the tests,

   Pr (rejecting the equality of the two means given that they are equal) $= \alpha$.

- Consequently,

   Pr (accepting the equality of the two means when they are equal) $= 1 - \alpha$.

- Therefore,

   Pr (accepting equality for all N, t tests performed) $= (1 - \alpha)^N$.

- Hence, finally we have

   Pr (rejecting the equality of *at least one pair of means* when H_0 is true)

   $= 1 - (1 - \alpha)^N (P$ say$)$.

- For particular values of k and for $\alpha = 0.05$, this results leads to the following numerical values:

k	N	P
2	1	$1 - (0.95)^1 = 0.05$
3	3	$1 - (0.95)^3 = 0.14$
4	6	$1 - (0.95)^6 = 0.26$
10	45	$1 - (0.95)^{45} = 0.90$

- The probability of falsely rejecting the null hypothesis quickly increases above the nominal significance level of .05. It is clear that such an approach is very likely to lead to misleading conclusions. Investigators unwise enough to apply the procedure would be led to claim more statistically significant results than their data justified.

significance level by almost threefold. The message is clear: avoid multiple t tests like the plague!

The appropriate method of analysis for a one-way design is the *one-way analysis of variance*.

3.3.   ONE-WAY ANALYSIS OF VARIANCE

The phrase "analysis of variance" was coined by arguably the most famous statistician of the twentieth century, Sir Ronald Aylmer Fisher, who defined it as "the separation of variance ascribable to one group of causes from the variance ascribable to the other groups." Stated another way, the analysis of variance (ANOVA) is a partitioning of the total variance in a set of data into a number of component parts, so that the relative contributions of identifiable sources of variation to the total variation in measured responses can be determined. But how does this separation or partitioning of *variance* help in assessing differences between *means*?

For the one-way design, the answer to this question can be found by considering two sample variances, one of which measures variation between the observations within the groups, and the other measures the variation between the group means. If the populations corresponding to the various levels of the independent variable have the same mean for the response variable, both the sample variances described estimate the *same* population value. However, if the population means differ, variation between the sample means will be greater than that between observations within groups, and therefore the two sample variances will be estimating *different* population values. Consequently, testing for the equality of the group means requires a test of the equality of the two variances. The appropriate procedure is an F test. (This explanation of how an F test for the equality of two variances leads to a test of the equality of a set of means is not specific to the one-way design; it applies to *all* ANOVA designs, which should be remembered in later chapters where the explanation is not repeated in detail.)

An alternative way of viewing the hypothesis test associated with the one-way design is that two alternative statistical models are being compared. In one the mean and standard deviation are the same in each population; in the other the means are different but the standard deviations are again the same. The F test assesses the plausibility of the first model. If the between group variability is greater than expected (with, say, $p < .05$), the second model is to be preferred.

A more formal account of both the model and the measures of variation behind the F tests in a one-way ANOVA is given in Display 3.2.

The data collected from a one-way design have to satisfy the following assumptions to make the F test involved strictly valid.

1. The observations in each group come from a normal distribution.
2. The population variances of each group are the same.
3. The observations are independent of one another.

Taking the third of these assumptions first, it is likely that in most experiments and investigations the independence or otherwise of the observations will be clear cut. When *unrelated* subjects are randomly assigned to treatment groups, for example, then independence clearly holds. And when the *same* subjects are observed under a number of different conditions, independence of observations is clearly unlikely, a situation we shall deal with in Chapter 5. More problematic are situations in which the groups to be compared contain possibly related subjects, for example, pupils within different schools. Such data generally require special techniques such as *multilevel modeling* for their analysis (see Goldstein, 1995, for details).

Display 3.2
The One-Way Analysis of Variance Model

- In a general sense, the usual model considered for a one-way design is that the subjects in a particular group come from a population with a particular *expected* or *average* value for the response variable, with differences between the subjects within a group being accounted for by some type of random variation, usually referred to as "error." So the model can be written as

$$\text{observed response} = \text{expected response} + \text{error}.$$

- The expected value in the population giving rise to the observations in the ith group is assumed to be μ_i, leading to the following model for the observations:

$$y_{ij} = \mu_i + \epsilon_{ij},$$

where y_{ij} represents the jth observation in the ith group, and the ϵ_{ij} represent random error terms, assumed to be from a normal distribution with mean zero and variance σ^2.

- The hypothesis of the equality of population means can now be written as

$$H_0 : \mu_1 = \mu_2 = \cdots = \mu_k = \mu,$$

leading to a new model for the observations, namely,

$$y_{ij} = \mu + \epsilon_{ij}.$$

- There are some advantages (and, unfortunately, some disadvantages) in reformulating the model slightly, by modeling the mean value for a particular population as the sum of the overall mean value of the response plus a specific population or group effect. This leads to a *linear model* of the form

$$y_{ij} = \mu + \alpha_i + \epsilon_{ij},$$

where μ represents the overall mean of the response variable, α_i is the effect on an observation of being in the ith group ($i = 1, 2, \ldots, k$), and again ϵ_{ij} is a random error term, assumed to be from a normal distribution with mean zero and variance σ^2.

- When written in this way, the model uses $k + 1$ parameters ($\mu, \alpha_1, \alpha_2, \ldots, \alpha_k$) to describe only k group means. In technical terms the model is said to be *overparameterized*, which causes problems because it is impossible to find unique estimates for each parameter—it's a bit like trying to solve simultaneous equations when there are fewer equations than unknowns.

- This feature of the usual form of ANOVA models is discussed in detail in Maxwell and Delaney (1990), and also briefly in Chapter 6 of this book. One way of overcoming the difficulty is to introduce the constraint $\sum_{i=1}^{k} \alpha_i = 0$, so that $\alpha_k = -(\alpha_1 + \alpha_2 + \cdots + \alpha_{k-1})$. The number of parameters is then reduced by one as required. (Other possible constraints are discussed in the exercises in Chapter 6.)

- If this model is assumed, the hypothesis of the equality of population means can be rewritten in terms of the parameters α_i as

$$H_0 : \alpha_1 = \alpha_2 = \cdots = \alpha_k = 0,$$

so that under H_0 the model assumed for the observations is

$$y_{ij} = \mu + \epsilon_{ij}$$

as before.

(Continued)

Display 3.2

(Continued)

- The total variation in the observations, that is, the sum of squares of deviations of observations from the overall mean of the response variable $[\sum_{i=1}^{k}\sum_{j=1}^{n}(y_{ij} - \bar{y}_{..})^2]$, can be partioned into that due to difference in group means, the *between groups sum of squares* $[\sum_{i=1}^{k} n(\bar{y}_{i.} - \bar{y}_{..})^2]$, where n is the number of observations in each group, and that due to differences among observations within groups, the *within groups sum of squares* $[\sum_{i=1}^{k}\sum_{j=1}^{n}(y_{ij} - \bar{y}_{i.})^2]$. (Here we have written out the formulas for the various sums of squares in mathematical terms by using the conventional "dot" notation, in which a dot represents summation over a particular suffix. In future chapters dealing with more complex ANOVA situations, we shall not bother to give the explicit formulas for sums of squares.)
- These sums of squares can be converted into between groups and within groups variances by dividing them by their appropriate degrees of freedom (see following text). Under the hypothesis of the equality of population means, both are estimates of σ^2. Consequently, an F test of the equality of the two variances is also a test of the equality of population means.
- The necessary terms for the F test are usually arranged in an *analysis of variance table* as follows ($N = kn$ is the total number of observations).

Source	DF	SS	MS	MSR(F)
Between groups	$k-1$	BGSS	BGSS$/(k-1)$	MSBG/MSWG
Within groups (error)	$N-k$	WGSS	WGSS$/(N-k)$	
Total	$N-1$			

Here, DF is degrees of freedom; SS is sum of squares; MS is mean square; BGSS is between groups sum of squares; and WGSS is within group sum of squares.
- If H_0 is true (and the assumptions discussed in the text are valid), the mean square ratio (MSR) has an F distribution with $k-1$ and $N-k$ degrees of freedom.
- Although we have assumed an equal number of observations, n, in each group in this account, this is not necessary; unequal group sizes are perfectly acceptable, although see the relevant comments made in the Summary section of the chapter.

The assumptions of normality and homogeneity of variance can be assessed by informal (graphical) methods and also by more formal approaches, such as the *Schapiro–Wilk test for normality* (Schapiro and Wilk, 1965) and *Bartlett's test for homogeneity of variance* (Everitt, 1998). My recommendation is to use the informal (which will be illustrated in the next section) and to largely ignore the formal. The reason for this recommendation is that formal tests of assumptions of both normality and homogeneity are of little practical relevance, because the good news is that even when the population variances are a little unequal and the observations are a little nonnormal, the usual F test is unlikely to mislead; the test is *robust* against minor departures from both normality and homogeneity of variance, particularly when the numbers of observations in each of the groups being compared are equal or approximately equal. Consequently, the computed p values will not be greatly distorted, and inappropriate conclusions are unlikely. Only if the departures from either or both the normality and homogeneity assumptions are extreme will there be real cause for concern, and a need to consider alternative procedures. Such gross departures will generally be visible from appropriate graphical displays of the data. When there *is* convincing evidence that the usual ANOVA approach should not be used, there are at least three possible alternatives.

1. *Transform* the raw data to make it more suitable for the usual analysis; that is, perform the analysis not on the raw data, but on the values obtained after applying some suitable mathematical function. For example, if the data are very skewed, a logarithm transformation may help. Transformations are discussed in detail by Howell (1992) and are also involved in some of the exercises in this chapter and in Chapter 4. Transforming the data is sometimes felt to be a trick used by statisticians, a belief that is based on the idea that the natural scale of measurement is sacrosanct in some way. This is not really the case, and indeed some measurements (e.g., pH values) are effectively already logarithmically transformed values. However, it is almost always preferable to present results in the original scale of measurement. (And it should be pointed out that these days there is perhaps not so much reason for psychologists to agonize over whether they should transform their data so as to meet assumptions of normality, etc., for reasons that will be made clear in Chapter 10.)

2. Use *distribution-free methods* (see Chapter 8).

3. Use a model that explicitly specifies more realistic assumptions than normality and homogeneity of variance (again, see Chapter 10).

3.4. ONE-WAY ANALYSIS OF VARIANCE OF THE FRUIT FLY FECUNDITY DATA

It is always wise to precede formal analyses of a data set with preliminary informal, usually graphical, methods. Many possible graphical displays were described in Chapter 2. Here we shall use box plots and probability plots to make an initial examination of the fruit fly data.

The box plots of the per diem fecundities of the three strains of fruit fly are shown in Figure 3.1. There is some evidence of an outlier in the resistant strain (per diem fecundity of 44), and also a degree of skewness of the observations in the susceptible strain, which may be a problem. More about the distributional properties of each strain's observations can be gleaned from the normal probability plots shown in Figure 3.2. Those for the resistant and nonselected strain suggest that normality is not an unreasonable assumption, but that the assumption is, perhaps, more doubtful for the susceptible strain. For now, however, we shall conveniently ignore these possible problems and proceed with an analysis of

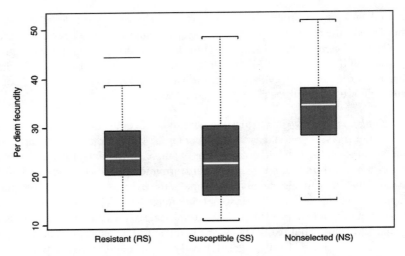

FIG. 3.1. Box plots of per diem fecundities for three strains of fruit fly.

variance of the raw data. (A far better way of testing the normality assumption in an ANOVA involves the use of quantities known as *residuals*, to be described in Chapter 4.)

The ANOVA table from a one-way analysis of variance of the fruit fly data is shown in Table 3.2. The F test for equality of mean fecundity in the three strains has an associated p value that is very small, and so we can conclude that the three strains do have different mean fecundities.

Having concluded that the strains do differ in their average fecundity does not, of course, necessarily imply that all strain means differ—the equality of means hypothesis might be rejected because the mean of one strain differs from the means of the other two strains, which are themselves equal. Discovering more about which particular means differ in a one-way design generally requires the use of what are known as *multiple comparison techniques*.

3.5. MULTIPLE COMPARISON TECHNIQUES

When a significant result has been obtained from a one-way analysis of variance, further analyses may be undertaken to find out more details of which particular means differ. A variety of procedures known generically as *multiple comparison techniques* are now available. These procedures all have the aim of retaining the nominal significance level at the required value when undertaking multiple tests of

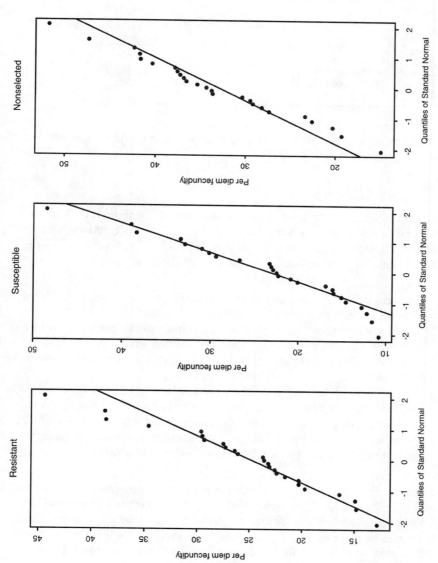

FIG. 3.2. Normal probability plots for per diem fecundities for three strains of fruit fly.

71

TABLE 3.2
One-Way ANOVA Results for Fruit Fly Data

Source	SS	DF	MS	F	p
Between strains	1362.21	2	681.11	8.67	<.001
Within strains (Error)	5659.02	72	78.60		

mean differences. We shall look at two multiple comparison procedures, the *Bonferonni test* and *Scheffé's test*.

3.5.1. Bonferonni Test

The essential feature of the Bonferonni approach to multiple testing is to compare each pair of means in turn, by using a Student's *t* test. However, is this not the very process dismissed as hopelessly misleading in Section 3.2? Well, yes and no is the somewhat unhelpful answer. Clearly some more explanation is needed.

1. The series of *t* tests contemplated here is carried out only after a significant *F* value is found in the one-way ANOVA.
2. Each *t* test here uses the pooled estimate of error variance obtained from the observations in all groups rather than just the two whose means are being compared. This estimate is simply the within groups mean square as defined in Display 3.2.
3. The problem of inflating the Type I error, as discussed in Section 3.2, is tackled by judging the *p* value from each *t* test against a significance level of α/m, where m is the number of *t* tests performed and α is the size of the Type I error (the significance level) that the investigator wishes to maintain in the testing process. (This is known as the *Bonferonni correction*.)

More details of the Bonferonni multiple comparison procedure are given in Display 3.3. The practical consequence of using this procedure is that each *t* test will now have to result in a more extreme value than usual for a significant mean difference to be claimed. In this way the overall Type I error will remain close to the desired value α. The disadvantage of the Bonferonni approach is that it may be highly conservative if a large number of comparisons are involved; that is, some real differences are very likely to be missed. (Contemplating a large number of comparisons may, of course, reflect a poorly designed study.) However, the procedure can be recommended for a small number of comparisons, and the results of its application to the fruit fly data are shown in Table 3.3. The results are

Display 3.3
Bonferroni t Tests

- The t statistic used is

$$t = \frac{\text{difference in group means}}{s[1/n_1 + 1/n_2]^{1/2}},$$

  where s^2 is the error mean square (an estimate of σ^2) and n_1 and n_2 are the number of observations in the two groups being compared.
- Each observed t statistic is compared with the value from a t distribution with $N - k$ degrees of freedom corresponding to a significance level of α/m rather than α, where m is the number of comparisons made.
- Alternatively (and preferably) a confidence interval can be constructed as

$$\text{mean difference} \pm t_{N-k}(\alpha/2m) \times s\sqrt{1/n_1 + 1/n_2},$$

  where $t_{N-k}(\alpha/2m)$ is the t value with $N - k$ degrees of freedom, corresponding to significance level $\alpha/2m$. (These confidence intervals are readily available from most statistical software; see computer hints at the end of the chapter.)

TABLE 3.3
Results from the Bonferonni Procedure Used on the Fruit Fly Data

- In this example $N - k = 72$ and $n_1 = n_2 = 25$ for each pair of groups. Three comparisons can be made between the pairs of means of the three strains, so that the t value corresponding to 72 degrees of freedom and $\alpha/6$ for $\alpha = 0.05$ is 2.45. Finally $s^2 = 78.60$ (see Table 3.2).
- The confidence intervals calculated as described in Display 3.3 are

Comparison	Est. of mean difference	Lower bound	Upper bound
NS–RS	8.12	1.97	14.30[a]
NS–SS	9.74	3.60	15.90[a]
RS–SS	1.63	−4.52	7.77

[a] Interval does not contain the value zero.

also shown in graphical form in Figure 3.3. Here it is clear that the mean fecundity of the nonselected strain differs from the means of the other two groups, which themselves do not differ.

3.5.2. Scheffé's Multiple Comparison Procedure

Scheffé's approach is particularly useful when a large number of comparisons are to be carried out. The test is again based on a Student's t test, but the critical point used for testing each of the t statistics differs from that used in the Bonferonni procedure. Details of the Scheffé test are given in Display 3.4, and the results of applying the test to the fruit fly data are shown in Table 3.4 and Figure 3.4.

TABLE 3.4
Scheffé's Procedure Applied to Fruit Fly Data

- Here the critical value is 2.50 and the confidence intervals for each comparison are

Comparison	Est. of mean difference	Lower bound	Upper bound
NS–RS	8.12	1.85	14.4[a]
NS–SS	9.74	3.48	16.0[a]
RS–SS	1.63	−4.64	7.9

[a]Interval does not contain the value zero.

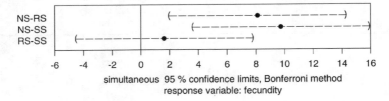

FIG. 3.3. Graphical display of Bonferroni multiple comparison results for fruit fly data.

Display 3.4
Scheffé's Multiple Comparison Procedure

- The test statistic used here is once again the t statistic used in the Bonferonni procedure and described in Display 3.3. In this case, however, each observed test statistic is compared with $[(k-1)F_{k-1, N-k}(\alpha)]^{1/2}$, where $F_{k-1, N-k}(\alpha)$ is the F value with $k-1$, $N-k$ degrees of freedom corresponding to significance level α. (More details are given in Maxwell and Delaney, 1990.)
- The method can again be used to construct a confidence interval for two means as

$$\text{mean difference} \pm \text{critical value} \times s\sqrt{1/n_1 + 1/n_2}.$$

The confidence intervals are very similar to those arrived at by the Bonferonni approach, and the conclusions are identical.

One general point to make about multiple comparison tests is that, on the whole, they all err on the side of safety (nonsignificance). Consequently, it is quite possible (although always disconcerting) to find that, although the F test in the analysis of variance is statistically significant, no pair of means is judged by the multiple comparison procedure to be significantly different. One further point is that a host of such tests are available, and statisticians are usually very wary of an overreliance

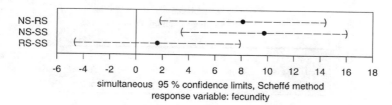

FIG. 3.4. Graphical display of Scheffé multiple comparison results for fruit fly data.

on such tests (the author is no exception). Readers need to avoid being seduced by the ease with which multiple comparison tests can be applied in most statistical software packages.

3.6. PLANNED COMPARISONS

Most analyses of one-way designs performed in psychological research involve the approach described in the previous two sections, namely a one-way ANOVA in which finding a significant F value is followed by the application of some type of multiple comparison test. But there is another way! Consider, for example, the data shown in Table 3.5, which were obtained from an investigation into the effect of the stimulant caffeine on the performance of a simple task. Forty male students were trained in finger tapping. They were then divided at random into four groups of 10, and the groups received different doses of caffeine (0, 100, 200, and 300 ml). Two hours after treatment, each man was required to carry out finger tapping and the number of taps per minute was recorded. Here, because the question of interest is whether caffeine affects performance on the finger-tapping task, the investigator may be interested *a priori* in the specific hypothesis that the mean of the three groups treated with caffeine differs from the mean of the untreated group, rather than in the general hypothesis of equality of means that is usually tested in a one-way analysis of variance. Such *a priori* planned comparisons are generally more powerful, that is, more likely to reject the null hypothesis when it is false, than the usual catch-all F test. The relevant test statistic for the specific hypothesis of interest can be constructed relatively simply (see Display 3.5). As also shown in Display 3.5, an appropriate confidence interval can be constructed for the comparison of interest. In the caffeine example, the interval constructed indicates that it is very likely that there is a difference in finger-tapping performance between the "no caffeine" and "caffeine" conditions. More finger tapping takes place when subjects are given the stimulant.

The essential difference between planned and unplanned comparisons (i.e., those discussed in the previous section) is that the former can be assessed by

TABLE 3.5
Caffeine and Finger-Tapping Data

0 ml	100 ml	200 ml	300 ml
242	248	246	248
245	246	248	250
244	245	250	251
248	247	252	251
247	248	248	248
248	250	250	251
242	247	252	252
244	246	248	249
246	243	245	253
242	244	250	251

Note. The response variable is the number of taps per minute.

using conventional significance levels whereas the latter require rather stringent significance levels. An additional difference is that when a planned comparison approach is used, omnibus analysis of variance is not required. The investigator moves straight to the comparisons of most interest. However, there is one caveat: planned comparisons have to be just that, and not the result of hindsight after inspection of the sample means!

3.7. THE USE OF ORTHOGONAL POLYNOMIALS: TREND ANALYSIS

In a one-way design where the independent variable is nominal, as in the teaching methods example, the data analysis is usually limited to testing the overall null hypothesis of the equality of the group means and subsequent *post hoc* comparisons by using some type of multiple comparison procedure. However, if the independent variable has levels that form an ordered scale, it is often possible to extend the analysis to examine the relationship of these levels to the group means of the dependent variable. An example is provided by the caffeine data in Table 3.5, where the levels of the independent variable take the values of 0 ml, 100 ml, 200 ml, and 300 ml. For such data most interest will center on the presence of *trends* of particular types, that is, increases or decreases in the means of the dependent variable over the ordered levels of the independent variable. Such

Display 3.5

Planned Comparisons

- The hypothesis of particular interest in the finger-tapping experiment is

$$H_0 : \mu_0 = 1/3[\mu_{100} + \mu_{200} + \mu_{300}],$$

where μ_0, μ_{100}, μ_{200}, and μ_{300} are the population means of the 0 ml, 100 ml, 200 ml, and 300 ml groups, respectively.
- The hypothesis can be tested by using the following t statistic:

$$t = \frac{\bar{x}_0 - 1/3(\bar{x}_{100} + \bar{x}_{200} + \bar{x}_{300})}{s(1/10 + 1/30)^{1/2}},$$

where $\bar{x}_0$, $\bar{x}_{100}$, $\bar{x}_{200}$, and $\bar{x}_{300}$ are the sample means of the four groups, and s^2 is once again the error mean square from the one-way ANOVA.
- The values of the four sample means are $\bar{x}_0 = 244.8$, $\bar{x}_{100} = 246.4$, $\bar{x}_{200} = 248.3$, and $\bar{x}_{200} = 250.4$. Here $s^2 = 4.40$.
- The t statistic takes the value -4.66. This is tested as a Student's t with 36 degrees of freedom (the degrees of freedom of the error mean square). The associated p value is very small ($p = .000043$), and the conclusion is that the tapping mean in the caffeine condition does not equal that when no caffeine is given.
- A corresponding 95% confidence interval for the mean difference between the two conditions can be constructed in the usual way to give the result $(-5.10, -2.04)$. More finger tapping occurs when caffeine is given.
- An equivalent method of looking at planned comparisons is to first put the hypothesis of interest in a slightly different form from that given earlier:

$$H_0 : \mu_0 - \tfrac{1}{3}\mu_{100} - \tfrac{1}{3}\mu_{200} - \tfrac{1}{3}\mu_{300} = 0.$$

- The estimate of this comparison of the four means (called a *contrast* because the defining constants, 1, $-1/3$, $-1/3$, $-1/3$ sum to zero), obtained from the sample means, is

$$244.8 - 1/3 \times 246.4 - 1/3 \times 248.3 - 1/3 \times 250.4 = -3.57.$$

- The sum of squares (and the mean square, because only a single degree of freedom is involved) corresponding to this comparison is found simply as

$$\frac{(-3.57)^2}{1/10[1 + (1/3)^2 + (1/3)^2 + (1/3)^2]} = 95.41.$$

(For more details, including the general case, see Rosenthal and Rosnow, 1985.)
- This mean square is tested as usual against the error mean square as an F with 1 and ν degrees of freedom, where ν is the number of degrees of freedom of the error mean square (in this case the value 36). So

$$F = 95.41/4.40 = 21.684.$$

The associated p value is .000043, agreeing with that of the t test described above.
- The two approaches outlined are exactly equivalent, because the calculated F statistic is actually the square of the t statistic ($21.68 = [-4.67]^2$).
- The second version of assessing a particular contrast by using the F statistic, and so on, will, however, help in the explanation of orthogonal polynomials to be given in Display 3.6.

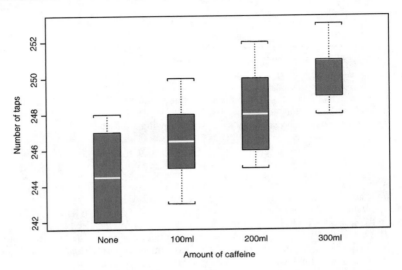

FIG. 3.5.   Box plots for finger-tapping data.

trends can be assessed relatively simply by using what are known as *orthogonal polynomials*. These correspond to particular comparisons among the levels of the independent variable, with the coefficients defining the required comparisons dependent on the number of levels of this variable. How these coefficients arise is not of great importance (and anyway is outside the level of this book), but how they are used *is* of interest; this is explained in Display 3.6. (Comprehensive tables of orthogonal polynomial coefficients corresponding to a number of different levels of the factor in a one-way design can be found in Howell, 1992.) Using these coefficients enables the between groups sum of squares to be partitioned into a number of single-degree-of-freedom terms, each of which represents the sum of squares corresponding to a particular trend component. (The arithmetic involved is similar to that described for planned comparisons in Display 3.4.) The particular results for the caffeine example are also shown in Display 3.6 and indicate that there is a very strong linear trend in the finger-tapping means over the four levels of caffeine. A box plot of the data demonstrates the effect very clearly (Figure 3.5).

3.8.   INTRODUCING ANALYSIS OF COVARIANCE

The data shown in Table 3.6 were collected from a sample of 24 first-grade children. Each child completed the Embedded Figures Test (EFT), which measures field dependence, that is, the extent to which a person can abstract the logical structure of a problem from its context. Then the children were randomly allocated to one

Display 3.6
The Use of Orthogonal Polynomials

- When the levels of the independent variable form a series of ordered steps, it is often of interest to examine the relationship between the levels and the means of the response variable. In particular the following questions would be of interest.
  1. Do the means of the treatment groups increase in a linear fashion with an increase in the level of the independent variable?
  2. Is the trend just linear or is there any evidence of nonlinearity?
  3. If the trend is nonlinear, what degree equation (polynomial) is required?
- The simplest way to approach these and similar questions is by the use of *orthogonal polynomial contrasts*. Essentially, these correspond to particular comparisons among the means, representing linear trend, quadratic trend, and so on. They are defined by a series of coefficients specific for the particular number of levels of the independent variable. The coefficients are available from most statistical tables. A small part of such a table is shown here.

k	Trend	Level 1	2	3	4	5
3	Linear	-1	0	1		
	Quadratic	1	-2	1		
4	Linear	-3	-1	1	3	
	Quadratic	1	-1	-1	1	
	Cubic	-1	3	-3	1	
5	Linear	-2	-1	0	1	2
	Quadratic	2	-1	-2	-1	2
	Cubic	-1	2	0	-2	1
	Quartic	1	-4	6	-4	1

(Note that these coefficients are only appropriate for equally spaced levels of the independent variable.)

- These coefficients can be used to produce a partition of the between groups sums of squares into single-degree-of-freedom sums of squares corresponding to the trend components. These sums of squares are found by using the approach described for planned comparisons in Display 3.5. For example, the sum of squares corresponding to the linear trend for the caffeine data is found as follows:

$$SS_{linear} = \frac{[-3 \times 244.8 + (-1) \times 246.4 + 1 \times 248.3 + 3 \times 250.4]^2}{1/10[(-3)^2 + (-1)^2 + (1)^2 + (3)^2]}$$

$$= 174.83.$$

- The resulting ANOVA table for the finger-tapping example is as follows.

Source	DF	SS	MS	F	p
Caffeine Levels	3	175.47	58.49	13.29	<.001
Linear	1	174.83	174.83	39.71	<.001
Quadratic	1	0.62	0.62	0.14	.71
Cubic	1	0.00	0.00	0.00	.97
Within	36	158.50	4.40		

- Note that the sum of squares of linear, quadratic, and cubic effects add to the between groups sum of squares.
- Note also that in this example the difference in the means of the four ordered caffeine levels is dominated by the *linear effect*; departures from linearity have a sum of squares of only $(0.62 + 0.00) = 0.62$.

TABLE 3.6
WISC Blocks Data

Row Group		Corner Group	
Time	EFT	Time	EFT
317	59	342	48
464	33	222	23
525	49	219	9
298	69	513	128
491	65	295	44
196	26	285	49
268	29	408	87
372	62	543	43
370	31	298	55
739	139	494	58
430	74	317	113
410	31	407	7

of two experimental groups. They were timed as they constructed a 3×3 pattern from nine colored blocks, taken from the Wechsler Intelligence Scale for Children (WISC). The two groups differed in the instructions they were given for the task: the "row group" were told to start with a row of three blocks, and the "corner group" were told to begin with a corner of three blocks. The experimenter was interested in whether the different instructions produced any change in the average time to complete the pattern, taking into account the varying field dependence of the children.

These data will help us to introduce a technique known as *analysis of covariance*, which allows comparison of group means after "adjusting" for what are called *concomitant variables,* or, more generally, *covariates.* The analysis of covariance tests for group mean differences in the response variable, after allowing for the possible effect of the covariate. The covariate is not in itself of experimental interest, except in that using it in an analysis can lead to increased precision by decreasing the estimate of experimental error, that is, the within group mean square in the analysis of variance. The effect of the covariate is allowed for by assuming a *linear* relationship between the response variable and the covariate. Details of the analysis of covariance model are given in Display 3.7. As described, the analysis of covariance assumes that the slopes of the lines relating response and covariate are the same in each group; that is, there is no interaction between the groups and the covariate. If there is an interaction, then it does not make sense to compare the groups at a single value of the covariate, because any difference noted will not apply for other values of the covariate.

Display 3.7
The Analysis of Covariance Model

- In general terms the model is

  observation = mean + group effect + covariate effect + error.

- More specifically, if y_{ij} is used to denote the value of the response variable for the jth individual in the ith group and x_{ij} is the value of the covariate for this individual, then the model assumed is

$$y_{ij} = \mu + \alpha_i + \beta(x_{ij} - \bar{x}) + \epsilon_{ij},$$

  where β is the regression coefficient linking response variable and covariate and $\bar{x}$ is the grand mean of the covariate values. (The remaining terms in the equation are as in the model defined in Display 3.2).

- Note that the regression coefficient is assumed to be the same in each group.
- The means of the response variable adjusted for the covariate are obtained simply as

  adjusted group mean = group mean + $\hat{\beta}$(grand mean of covariate

  − group mean of covariate),

  where $\hat{\beta}$ is the estimate of the regression coefficient in the model above.

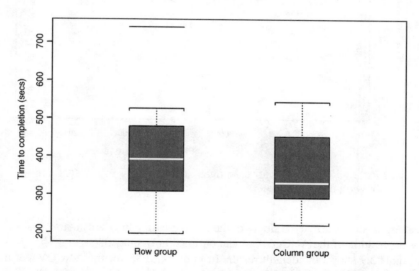

FIG. 3.6. Box plots of completion times for the WISC data.

Before applying analysis of covariance (ANCOVA) to the WISC data, we should look at some graphical displays of the data. Box plots of the recorded completion times in each experimental group are shown in Figure 3.6, and a scattergram giving the fitted regressions (see Chapter 6) for time against the EFT in each group is given in Figure 3.7. The box plots suggest a possible outlier. Here we shall conveniently ignore this possible problem and analyze all the data (but see Exercise 3.5). The

TABLE 3.7
ANCOVA Results from SPSS for WISC Blocks Data

Source	SS	DF	MS	F	p
Field dependence	109991	1	109991	9.41	.006
Group	11743	1	11743	1.00	.33
Error	245531	21	11692		

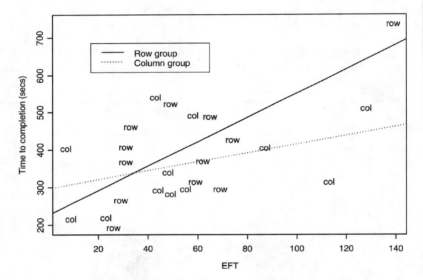

FIG. 3.7. Scatterplot for time against the EFT for the WISC data, showing fitted regressions for each group.

scatterplot suggests that the slopes of the two regression lines are not too dissimilar for application of the ANCOVA model outlined in Display 3.7.

Table 3.7 shows the default results from using SPSS for the ANCOVA of the data in Table 3.6. In Table 3.8 the corresponding results from using S-PLUS for the analysis are shown. You will notice that the results are similar but not identical The test for the covariate differs. The reason for this difference will be made clear in Chapter 4. However despite the small difference in the results from the two packages, the conclusion from each is the same: it appears that field dependence is predictive of time to completion of the task, but that there is no difference between the two experimental groups in mean completion time.

TABLE 3.8
ANCOVA of WISC Data by Using S-PLUS

Source	SS	DF	MS	F	p
Field dependence	110263	1	110263	9.43	0.006
Group	11743	1	11743	1.00	0.33
Error	245531	21	11692		

Further examples of analysis of covariance are given in the next chapter, and then in Chapter 6 the model is put in a more general context. However, one point about the method that should be made here concerns its use with naturally occurring or intact groups, rather than groups formed by random allocation.

The analysis of covariance was originally intended to be used in investigations in which randomization had been used to assign patients to treatment groups. Experimental precision could be increased by removing from the error term in the ANOVA that part of the residual variability in the response that was linearly predictable from the covariate. Gradually, however, the technique became more widely used to test hypotheses that were generally stated in such terms as "the mean group differences on the response are zero when the group means on the covariate are made equal," or "the group means on the response after adjustment for mean differences on the covariate are equal." Indeed, some authors (e.g., McNemar, 1962) have suggested that "if there is only a small chance difference between the groups on the covariate, the use of covariance adjustment may not be worth the effort." Such a comment rules out the very situation for which the ANCOVA was originally intended, because in the case of groups formed by randomization, any group differences on the covariate are necessarily the result of chance! Such advice is clearly unsound because more powerful tests of group differences result from the decrease in experimental error achieved when analysis of covariance is used in association with random allocation.

In fact, it is the use of analysis of covariance in an attempt to undo built-in differences among intact groups that causes concern. For example, Figure 3.8 shows a plot of reaction time and age for psychiatric patients belonging to two distinct diagnostic groups. An ANCOVA with reaction time as response and age as covariate might lead to the conclusion that reaction time does not differ in the two groups. In other words, given that two patients, one from each group, are of approximately the same age, their reaction times are also likely to be similar. Is such a conclusion sensible? An examination of Figure 3.6 clearly shows that it is not, because the ages of the two groups do not overlap and an ANOVA has

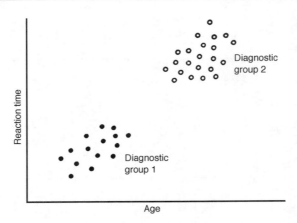

FIG. 3.8.   Plot of reaction time against age for two groups of psychiatric patients.

essentially extrapolated into a region with no data. Presumably, it is this type of problem that provoked the following somewhat acidic comment from Anderson (1963): "one may well wonder what exactly it means to ask what the data would be like if they were not like they are!" Clearly, some thought has to be given to the use of analysis of covariance on intact groups, and readers are referred to Fleiss and Tanur (1972) and Stevens (1992) for more details.

3.9. HOTELLING'S T^2 TEST AND ONE-WAY MULTIVARIATE ANALYSIS OF VARIANCE

The data shown in Table 3.9 were obtained from a study reported by Novince (1977), which was concerned with improving the social skills of college females and reducing their anxiety in heterosexual encounters. There were three groups in the study: a control group, a behavioral rehearsal group, and a behavioral rehearsal plus cognitive restructuring group. The values of the following four dependent variables were recorded for each subject in the study:

- anxiety–physiological anxiety in a series of heterosexual encounters;
- measure of social skills in social interactions;
- appropriateness;
- assertiveness.

Between group differences in this example could be assessed by separate one-way analyses of variance on each of the four dependent variables. An alternative

TABLE 3.9
Social Skills Data

Anxiety	Social Skills	Appropriateness	Assertiveness
Behavioral Rehearsal			
5	3	3	3
5	4	4	3
4	5	4	4
4	5	5	4
3	5	5	5
4	5	4	4
4	5	5	5
4	4	4	4
5	4	4	3
5	4	4	3
4	4	4	4
Control Group			
6	2	1	1
6	2	2	2
5	2	3	3
6	2	2	2
4	4	4	4
7	1	1	1
5	4	3	3
5	2	3	3
5	3	3	3
5	4	3	3
6	2	3	3
Behavioral Rehearsal + Cognitive Restructuring			
4	4	4	4
4	3	4	3
4	4	4	4
4	5	5	5
4	5	5	5
4	4	4	4
4	5	4	4
4	6	6	5
4	4	4	4
5	3	3	3
4	4	4	4

Note. From Novince (1977).

and, in many situations involving multiple dependent variables, a preferable procedure is to consider the four dependent variables *simultaneously*. This is particularly sensible if the variables are correlated and believed to share a common conceptual meaning. In other words, the dependent variables considered together make sense as a group. Consequently, the real question of interest is, Does the set of variables as a whole indicate any between group differences? To answer this question requires the use of a relatively complex technique known as multivariate analysis of variance, or MANOVA. The ideas behind this approach are introduced most simply by considering first the two-group situation and the multivariate analog of Student's t test for testing the equality of two means, namely *Hotelling's T^2 test*. The test is described in Display 3.8 (the little adventure in matrix algebra is unavoidable, I'm afraid), and its application to the two experimental groups in Table 3.9 is detailed in Table 3.10. The conclusion to be drawn from the T^2 test is that the mean vectors of the two experimental groups do not differ.

It might be thought that the results from Hotelling's T^2 test would simply reflect those of a series of univariate t tests, in the sense that if no differences are found by the separate t tests, then Hotelling's T^2 test will also lead to the conclusion that the population multivariate mean vectors do not differ, whereas if any significant difference is found for the separate variables, the T^2 statistic will also be significant. In fact, this is not necessarily the case (if it were, the T^2 test would be a waste of time); it is possible to have no significant differences for each variable tested separately but a significant T^2 value, and vice versa. A full explanation of the differences between a univariate and multivariate test for a situation involving just two variables is given in Display 3.9.

When a set of dependent variables is to be compared in more than two groups, the multivariate analog of the one-way ANOVA is used. Without delving into the technical details, let us say that what the procedure attempts to do is to combine the dependent variables in some optimal way, taking into consideration the correlations between them and deciding what unique information each variable provides. A number of such composite variables giving maximum separation between groups are derived, and these form the basis of the comparisons made. Unfortunately, in the multivariate situation, when there are more than two groups to be compared, no single test statistic can be derived that is always the most powerful for detecting all types of departures from the null hypothesis of the equality of the mean vectors of the groups. A number of different test statistics have been suggested that may give different results when used on the same set of data, although the resulting conclusion from each is often the same. (When only two groups are involved, all the proposed test statistics are equivalent to Hotelling's T^2.) Details and formulas for the most commonly used MANOVA test statistics are given in Stevens (1992), for example, and in the glossary in Appendix A. (The various test statistics are eventually transformed into F statistics

Display 3.8

Hotelling's T^2 Test

- If there are p dependent variables, the null hypothesis is that the p means of the first population equal the corresponding means of the second population.
- By introduction of some vector and matrix nomenclature, the null hypothesis can be written as

$$H_0 : \boldsymbol{\mu}_1 = \boldsymbol{\mu}_2,$$

where $\boldsymbol{\mu}_1$ and $\boldsymbol{\mu}_2$ contain the population means of the dependent variables in the two groups; that is, they are the population *mean vectors*.
- The test statistic is

$$T^2 = \frac{n_1 n_2 D^2}{n_1 + n_2},$$

when n_1 and n_2 are the numbers of observations in the two groups and D^2 is defined as

$$D^2 = (\bar{\mathbf{x}}_1 - \bar{\mathbf{x}}_2)'\mathbf{S}^{-1}(\bar{\mathbf{x}}_1 - \bar{\mathbf{x}}_2),$$

where $\bar{\mathbf{x}}_1$ and $\bar{\mathbf{x}}_2$ are the sample mean vectors in the two groups and $\mathbf{S}$ is the estimate of the assumed common covariance matrix calculated as

$$\mathbf{S} = \frac{(n_1 - 1)\mathbf{S}_1 + (n_2 - 1)\mathbf{S}_2}{n_1 + n_2 - 2},$$

where $\mathbf{S}_1$ and $\mathbf{S}_2$ are the sample covariance matrices in each group.
- Note that the form of the multivariate test statistic is very similar to that of the univariate independent sample t test of your introductory course; it involves a difference between the "means" (here the means are vectors and the difference between them takes into account the covariances of the variables) and an assumed common "variance" (here the variance becomes a covariance matrix).
- Under H_0 (and when the assumptions listed below are true), the statistic F given by

$$F = \frac{(n_1 + n_2 - p - 1)T^2}{(n_1 + n_2 - 2)p}$$

has a Fisher F distribution with p and $n_1 + n_2 - p - 1$ degrees of freedom.
- The assumptions of the test are completely analogous to those of an independent samples t test. They are as follows: (a) the data in each population are from a multivariate normal distribution; (b) the populations have the same covariance matrix; and (c) the observations are independent.

---

to enable p values to be calculated.) Only the results of applying the tests to the data in Table 3.9 will be discussed here (the results are given in Table 3.11). Each test indicates that the three groups are significantly different on the set of four variables. Combined with the earlier T^2 test on the two experimental groups, this result seems to imply that it is the control group that gives rise to the difference. This

TABLE 3.10
Calculation of Hotelling's T^2 for the Data in Table 3.8

The two sample mean vectors are given by

$$\mu_1' = [4.27, 4.36, 4.18, 2.82], \quad \mu_2' = [4.09, 4.27, 4.27, 4.09].$$

The sample covariance matrices of the two groups are

$$S_1 = \begin{pmatrix} 0.42 & -0.31 & -0.25 & -0.44 \\ -0.31 & 0.45 & 0.33 & 0.37 \\ -0.25 & 0.33 & 0.36 & 0.34 \\ -0.44 & 0.37 & 0.34 & 0.56 \end{pmatrix},$$

$$S_2 = \begin{pmatrix} 0.09 & -0.13 & -0.13 & -0.11 \\ -0.13 & 0.82 & 0.62 & 0.57 \\ -0.13 & 0.62 & 0.62 & 0.47 \\ -0.11 & 0.57 & 0.47 & 0.49 \end{pmatrix}.$$

The combined covariance matrix is

$$S = \begin{pmatrix} 0.28 & -0.24 & -0.21 & -0.31 \\ -0.24 & 0.71 & 0.52 & 0.52 \\ -0.21 & 0.52 & 0.54 & 0.45 \\ -0.31 & 0.52 & 0.45 & 0.59 \end{pmatrix}.$$

Consequently, $D^2 = 0.6956$ and $T^2 = 3.826$.
The resulting F value is 0.8130 with 4 and 17 degrees of freedom.
The corresponding p value is .53.

Display 3.9
Univariate and Multivariate Tests for Equality of Means for Two Variables

- Suppose we have a sample of n observations on two variables x_1 and x_2, and we wish to test whether the population means of the two variables μ_1 and μ_2 are both zero.
- Assume the mean and standard deviation of the x_1 observations are $\bar{x}_1$ and s_1 respectively, and of the x_2 observations, $\bar{x}_2$ and s_2.
- If we test separately whether each mean takes the value zero, then we would use two t tests. For example, to test $\mu_1 = 0$ against $\mu_1 \neq 0$ the appropriate test statistic is

$$t = \frac{\bar{x}_1 - 0}{s_1/\sqrt{n}}.$$

- The hypothesis $\mu_1 = 0$ would be rejected at the α percent level of significance if $t < -t_{100(1-\frac{1}{2}\alpha)}$ or $t > t_{100(1-\frac{1}{2}\alpha)}$; that is, if $\bar{x}_1$ fell outside the interval $[-s_1 t_{100(1-\frac{1}{2}\alpha)}/\sqrt{n}, \ s_1 t_{100(1-\frac{1}{2}\alpha)}/\sqrt{n}]$ where $t_{100(1-\frac{1}{2}\alpha)}$ is the $100(1-\frac{1}{2}\alpha)$ percent point of the t distribution with $n-1$ degrees of freedom. Thus the hypothesis would *not* be rejected if $\bar{x}_1$ fell *within* this interval.

(Continued)

Display 3.9

(Continued)

- Similarly, the hypothesis $\mu_2 = 0$ for the variable x_2 would not be rejected if the mean, $\bar{x}_2$, of the x_2 observations fell within a corresponding interval with s_2 substituted for s_1.
- The multivariate hypothesis $[\mu_1, \mu_2] = [0, 0]$ would therefore not be rejected if *both* these conditions were satisfied.
- If we were to plot the point $(\bar{x}_1, \bar{x}_2)$ against rectangular axes, the area within which the point could lie and the multivariate hypothesis not rejected is given by the rectangle ABCD of the diagram below, where AB and DC are of length $2s_1 t_{100(1-\frac{1}{2}\alpha)}\sqrt{n}$ while AD and BC are of length $2s_2 t_{100(1-\frac{1}{2}\alpha)}\sqrt{n}$.

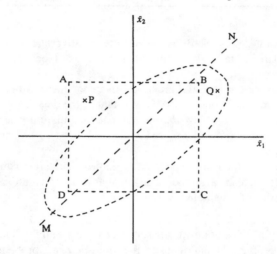

- Thus a sample that gave the means $(\bar{x}_1, \bar{x}_2)$ represented by the point P would lead to acceptance of the mutivariate hypothesis.
- Suppose, however, that the variables x_1 and x_2 are moderately highly correlated. Then all points (x_1, x_2) and hence $(\bar{x}_1, \bar{x}_2)$ should lie reasonably close to the straight line MN through the origin marked on the diagram.
- Hence samples consistent with the multivariate hypothesis should be represented by points $(\bar{x}_1, \bar{x}_2)$ that lie within a region encompassing the line MN. When we take account of the nature of the variation of bivariate normal samples that include correlation, this region can be shown to be an ellipse such as that marked on the diagram. The point P is *not* consistent with this region and, in fact, should be *rejected* for this sample.
- Thus the inference drawn from the two separate univariate tests conflicts with the one drawn from a single multivariate test, and it is the wrong inference.
- A sample giving the $(\bar{x}_1, \bar{x}_2)$ values represented by point Q would give the other type of mistake, where the application of two separate univariate tests leads to the rejection of the null hypothesis, but the correct multivariate inference is that the hypothesis should *not* be rejected. (This explanation is taken with permission from Krzanowski, 1991.)

TABLE 3.11

Multivariate Tests for the Data in Table 3.9

Test Name	Value	Corresponding F Value	DF 1	DF 2	p
Pillai	0.67980	3.60443	8.00	56.00	.002
Hotelling	1.57723	5.12600	8.00	52.00	<.001
Wilk	0.36906	4.36109	8.00	54.00	<.001

conclusion could be checked by using the multivariate equivalent of the multiple comparison tests described in Section 3.5 (see Stevens, 1992, for details).

(The multivariate test statistics are based on analogous assumptions to those of the F tests in a univariate one-way ANOVA; that is, the data are assumed to come from a *multivariate normal distribution* and each population is assumed to have the same *covariance matrix*. For more details see Stevens, 1992, and the glossary in Appendix A.)

In situations in which the dependent variables of interest can genuinely be regarded as a set, there are a number of advantages in using a multivariate approach rather than a series of separate univariate analyses.

1. The use of a series of univariate tests leads to a greatly inflated Type 1 error rate; the problem is analogous to the multiple t test problem described in Section 3.2.
2. The univariate tests ignore important information, namely the correlations among the dependent variables. The multivariate tests take this information into account.
3. Although the groups may not be significantly different on any of the variables individually, jointly the set of variables may reliably differentiate the groups; that is, small differences on several variables may combine to produce a reliable overall difference. Thus the multivariate test will be more powerful in this case.

However, it should be remembered that these advantages are genuine only if the dependent variables can honestly be regarded as a set sharing a conceptual meaning. A multivariate analysis of variance on a number of dependent variables when there is not strong rationale for regarding them simultaneously is *not* to be recommended.

3.10. SUMMARY

1. A one-way ANOVA is used to compare the means of k populations when $k \geq 3$.
2. The null and alternative hypotheses are

$$H_0 : \mu_1 = \mu_2 = \cdots = \mu_k,$$

$$H_1 : \text{not all means are equal.}$$

3. The test of the null hypothesis involves an F test for the equality of two variances.
4. A significant F test should be followed by one or other multiple comparison test to assess which particular means differ, although care is needed in the application of such tests.
5. When the levels of the independent variable form an ordered scale, the use of orthogonal polynomials is often informative.
6. Including variables that are linearly related to the response variable as co-variates increases the precision of the study in the sense of leading to a more powerful F test for the equality of means hypothesis.
7. The examples in this chapter have all involved the same number of subjects in each group. This is not a necessary condition for a one-way ANOVA, although it is helpful because it makes depatures from the assumptions of normality and homogeneity of variance even less critical than usual. Some of the formulas in displays require minor amendments for unequal group sizes.
8. In the one-way ANOVA model considered in this chapter, the levels of the independent variable have been regarded as *fixed*; that is, the levels used are those of specific interest. An alternative model in which the levels are considered as a *random sample* from some population of possible levels might also have been considered. However, because such models are usually of greater interest in the case of more complex designs, their description will be left until later chapters.
9. Several dependent variables can be treated simultaneously by MANOVA; this approach is only really sensible if the variables can truly be regarded, in some sense, as a unit.

COMPUTER HINTS

SPSS

In SPSS the first steps in conducting a one-way ANOVA are as follows.

1. Click **Statistics**, click **General Linear Model**, and then click **GLM-General Factorial**. The GLM-General Factorial dialog box will appear.

2. Select the response variable and the grouping variable.
3. Click on **Options** to choose any particular options required such as homogeneity tests, estimates of effect sizes, descriptive statistics, and so on.

To conduct a one-way MANOVA, the basic steps are are as follows.

1. Click **Statistics**, click **General Linear Model**, and then click **GLM-Multivariate**. The GLM-Multivariate dialog box now becomes available.
2. Select the dependent variables and the grouping variable from the relevant data set.
3. Click **Options** to refine the analysis and select things such as estimates and post hoc tests.

S-PLUS

In S-PLUS, a simple one-way ANOVA can be conducted by using the following steps.

1. Click **Statistics**, click **ANOVA**, and click **Fixed Effects**; then the ANOVA dialog box will appear.
2. Choose the relevant data set and select the dependent variable and the grouping (independent) variable.
3. Click the **Plot** tag of the dialog box if, for example, some type of residual plot is required.

A one-way MANOVA can be undertaken as follows.

1. Click **Statistics**, click **Multivariate**, and click **MANOVA** to get the MANOVA dialog.
2. Select the multiple response variables and the grouping variable.

Multiple comparison test are conducted by using the following.

1. Click **Statistics**, click **ANOVA**, and click **Multiple Comparisons** to get the Multiple Comparisons dialog.
2. Select the resulting analysis of variance object saved from a previous analysis of variance and select the method required, the confidence level, and so on.

In S-PLUS, an ANOVA can be undertaken by means of the command line approach using the *aov* function. If, for example, the fruit fly data (see Table 3.1) were stored in a data frame called *fruit* with variable names *group* and *fecund*, a one-way ANOVA could be carried out by using the command

$$aov(fecund \sim group, data = fruit).$$

Multiple comparisons can be applied by using the *multicomp* function. The multiple comparison procedure produces a useful graphical display of the results (see Figures 3.3 and 3.4).

A multivariate analysis of variance is available by using the *manova* function.

EXERCISES

3.1. Reproduce all the results given in this chapter by using your favorite piece of statistical software. (This exercise should be repeated in subsequent chapters, and those readers finding any differences between their results and mine are invited to e-mail me at *b.everitt@iop.kcl.ac.uk.*)

3.2. The data given in Table 3.12 were collected in an investigation described by Kapor (1981), in which the effect of knee-joint angle on the efficiency of cycling was studied. Efficiency was measured in terms of the distance pedalled on an ergocycle until exhaustion. The experimenter selected three knee-joint angles

TABLE 3.12
The Effect of Knee-Joint Angle on the Efficiency
of Cycling: Total Distance Covered (km)

50°	70°	90°
8.4	10.6	3.2
7.0	7.5	4.2
3.0	5.1	3.1
8.0	5.6	6.9
7.8	10.2	7.2
3.3	11.0	3.5
4.3	6.8	3.1
3.6	9.4	4.5
8.0	10.4	3.8
6.8	8.8	3.6

of particular interest: 50°, 70°, and 90°. Thirty subjects were available for the experiment and 10 subjects were randomly allocated to each angle. The drag of the ergocycle was kept constant at 14.7 N, and subjects were instructed to pedal at a constant speed of 20 km/h.

1. Carry out an initial graphical inspection of the data to assess whether there are any aspects of the observations that might be a cause for concern in later analyses.

2. Derive the appropriate analysis of variance table for the data.

3. Investigate the mean differences in more detail using a suitable multiple comparisons test.

3.3. Suggest suitable estimators for the parameters μ and α_i in the one-way ANOVA model given in Display 3.2. Use your suggested estimators on the fruit fly data.

3.4. The data in Table 3.13 were collected in an investigation of maternal behavior of laboratory rats. The response variable was the time (in seconds) required for the mother to retrieve the pup to the nest, after being moved a fixed distance away. In the study, pups of different ages were used.

1. Carry out a one way analysis of variance of the data.

2. Use an orthogonal polynomial approach to investigate whether there is any evidence of a linear or quadratic trend in the group means.

3.5. The data in Table 3.14 show the anxiety scores, both before and after the operation, for patients undergoing wisdom tooth extraction by one of three

TABLE 3.13
Maternal Behavior in Rats

	Age of Pup	
5 Days	20 Days	35 Days
15	30	40
10	15	35
25	20	50
15	25	43
20	23	45
18	20	40

TABLE 3.14
Anxiety Scores for Patients Having Wisdom Teeth Extracted

Method of Extraction	Age (years)	Initial Anxiety	Final Anxiety
Method 1	27	30.2	32.0
	32	35.3	34.8
	23	32.4	36.0
	28	31.9	34.2
	30	28.4	30.3
	35	30.5	33.2
	32	34.8	35.0
	21	32.5	34.0
	26	33.0	34.2
	27	29.9	31.1
Method 2	29	32.6	31.5
	29	33.0	32.9
	31	31.7	34.3
	36	34.0	32.8
	23	29.9	32.5
	26	32.2	32.9
	22	31.0	33.1
	20	32.0	30.4
	28	33.0	32.6
	32	31.1	32.8
Method 3	33	29.9	34.1
	35	30.0	34.1
	21	29.0	33.2
	28	30.0	33.0
	27	30.0	34.1
	23	29.6	31.0
	25	32.0	34.0
	26	31.0	34.0
	27	30.1	35.0
	29	31.0	36.0

methods. The ages of the patients (who were all women) were also recorded. Patients were randomly assigned to the three methods.

1. Carry out a one-way ANOVA on the anxiety scores on discharge.

2. Construct some scatterplots to assess informally whether the relationship between anxiety score on discharge and anxiety score prior to the operation is the same in each treatment group, and whether the relationship is linear.

3. If you think it is justified, carry out an ANCOVA of the anxiety scores on discharge, using anxiety scores prior to the operation as a covariate.

4. Carry out an ANOVA of the difference between the two anxiety scores.

5. Comment on the difference between the analyses in steps 3 and 4.

6. Suggest a suitable model for an ANCOVA of anxiety score on discharge by using the two covariates of anxiety score prior to the operation and age. Carry out the appropriate analysis.

3.6. Reanalyze the WISC data in Table 3.6 after removing any observation you feel might reasonably be labeled as an outlier. Calculate the adjusted group means for time to completion by using the formula given in Display 3.7.

3.7. Show that when only a single variable is involved, the test statistic for Hotelling's T^2 given in Display 3.8 is equivalent to the test statistic used in an independent samples t test.

3.8. Apply Hotelling's T^2 test to each pair of groups in the data in Table 3.9. Use the three T^2 values and the Bonferroni correction procedure to assess which groups differ.

4

Analysis of Variance II: Factorial Designs

4.1. INTRODUCTION

Many experiments in psychology involve the simultaneous study of the effects of two or more factors on a response variable. Such an arrangement is usually referred to as a *factorial design*. Consider, for example, the data shown in Table 4.1, taken from Lindman (1974), which gives the improvement scores made by patients with one of two diagnoses, being treated with different tranquillizers. The questions of interest about these data concern the equality or otherwise of the average improvement in the two diagostic categories, and similarly for the different drugs. Because these questions are the same as those considered in the previous chapter, the reader might enquire why we do not simply apply a one-way ANOVA separately to the data for diagostic categories and for drugs. The answer is that such analyses would omit an aspect of a factorial design that is often very important, as we shall demonstrate in the next section.

4.2. INTERACTIONS IN A FACTORIAL DESIGN

The results of a one-way analysis of (1) psychiatric categories only, (2) tranquillizer drugs only, and (3) all six groups of observations in Table 4.1 are shown in Table 4.2. At first sight the results appear a little curious. The F test for the equality of the means of the psychiatric category is not significant, and neither is that for the

97

TABLE 4.1
Improvement Scores for Two Psychiatric Categories
and Three Tranquillizer Drugs

		Drug	
Category	B1	B2	B3
A1	8	8	4
	4	10	6
	0	6	8
A2	10	0	15
	6	4	9
	14	2	12

TABLE 4.2
ANOVA of Improvement Scores in Table 4.1

Source	SS	DF	MS	F	p
Psychiatric Categories Only					
Diagosis	20.05	1	20.05	1.07	0.31
Error	289.90	16	18.68		
Drugs Only					
Drugs	44.11	2	22.05	1.20	0.33
Error	274.83	15	18.32		
All Six Groups					
Groups	212.28	5	42.45	4.78	0.01
Error	106.67	12	8.89		

tranquillizer drug means. Nevertheless, the *F* test for the six groups together *is* significant, indicating a difference between the six means that has not been detected by the separate one-way analyses of variance. A clue to the cause of the problem is provided by considering the degrees of freedom associated with each of the three between groups sum of squares. That corresponding to psychiatric categories has a single degree of freedom and that for drugs has two degrees of freedom, making a total of three degrees of freedom for the separate analyses. However, the between groups sum of squares for the six groups combined has five degrees of freedom.

The separate one-way analyses of variance appear to have omitted some aspect of the variation in the data. What has been left out is the effect of the combination of factor levels that is not predictable from the sum of the effects of the two factors separately, an effect usually known as the *interaction* between the factors. Both the model for, and the analysis of, a factorial design have to allow for the possibility of such interaction effects.

4.3. TWO-WAY DESIGNS

Many aspects of the modeling and analysis of factorial designs can be conveniently illustrated by using examples with only two factors. Such *two-way designs* are considered in this section. An example with more than two factors will be discussed in the next section.

The general model for a two-way design is described in Display 4.1. Notice that this model contains a term to represent the possible interaction between factors.

Display 4.1
Model for a Two-Way Design with Factors A and B

- The observed value of the response variable for each subject is assumed to be of the form

$$\text{observed response} = \text{expected response} + \text{error}.$$

- The expected value of the response variable for a subject is assumed to be made up of an effect for the level of A under which the subject is observed, the corresponding term for factor B, and the interaction effect for the particular AB combination. Consequently, the model above can be rewritten as

$$\text{observed response} = \text{overall population mean} +$$
$$\text{factor A effect} + \text{factor B effect}$$
$$+ \text{AB interaction effect} + \text{error}.$$

- More specifically, let y_{ijk} represent the kth observation in the jth level of factor B (with b levels) and the ith level of factor A (with a levels). Assume there are n subjects in each cell of the design. The levels of A and B are assumed to be of particular interest to the experimenter, and so both factors and their interaction are regarded as having fixed effects (see Section 4.5).
- The model assumed for the observations can now be written as

$$y_{ijk} = \mu + \alpha_i + \beta_j + \gamma_{ij} + \epsilon_{ijk},$$

where μ represents the overall mean, α_i is the effect on an observation of being in the ith level of factor A, β_j is the corresponding effect for the jth level of factor B, γ_{ij} represents the interaction effect, and ϵ_{ijk} is a random error term assumed to have a normal distribution with mean zero and variance σ^2.

(Continued)

Display 4.1
(Continued)

- Formulated in this way, the model has too many parameters and constraints have to be introduced as explained in Chapter 3. The most common method is to require that the parameters in this fixed-effects model are such that $\sum_{i=1}^{a} \alpha_i = \sum_{j=1}^{b} \beta_j = \sum_{i=1}^{a} \gamma_{ij} = \sum_{j=1}^{b} \gamma_{ij} = 0$. For further details, see Maxwell and Delaney (1990).
- The hypotheses of interest, namely no factor A effect, no factor B effect, and no interaction, imply the following about the parameters of the model:

$$
\begin{aligned}
H_0^{(A)} &: \alpha_1 = \alpha_2 = \cdots = \alpha_a = 0, \\
H_0^{(B)} &: \beta_1 = \beta_2 = \cdots = \beta_b = 0, \\
H_0^{(AB)} &: \gamma_{11} = \gamma_{12} = \cdots = \gamma_{ab} = 0.
\end{aligned}
$$

- Under each hypothesis, a different model is assumed to be adequate to describe the observations; for example, under $H_0^{(AB)}$ this model is

$$
y_{ijk} = \mu + \alpha_i + \beta_j + \epsilon_{ijk},
$$

that is, the *main effects model*.
- The total variation in the observation is partioned into that caused by differences between the levels of the factor A, that caused by differences between the levels of factor B, that caused by the interaction of A and B, and that caused by differences among observations in the same cell.
- Under the hull hypotheses given above, the factor A, the factor B, and the interaction mean squares are all estimators of σ^2. The error mean square is also an estimate of σ^2 but one that does not rely on the truth or otherwise of any of the null hypotheses. Consequently, each mean square ratio can be tested as an F statistic to assess each of the null hypotheses specified.
- The analysis of variance table for the model is

Source	SS	DF	MS	F
A	ASS	$a-1$	$\frac{ASS}{(a-1)}$	MSA/error MS
B	BSS	$b-1$	$\frac{BSS}{(b-1)}$	MSB/error MS
AB	ABSS	$(a-1)(b-1)$	$\frac{ABSS}{(a-1)(b-1)}$	MSAB/error MS
Error	WCSS	$ab(n-1)$	$\frac{WCSS}{ab(n-1)}$	
(within cells)				

- The F tests are valid under the following assumptions: (a) normality of the error terms in the model, (b) homogeneity of variance, and (c) independence of the observations for each subject.

The total variation in the observations can be partitioned into parts due to each factor and a part due to their interaction as shown.

The results of applying the model described in Display 4.1 to the psychiatric data in Table 4.1 are shown in Table 4.3. The significant interaction term now explains the previously confusing results of the separate one-way analyses of variance. (The detailed interpretation of interaction effects will be discussed later.)

TABLE 4.3
Two-Way ANOVA of Psychiatric Improvement Scores

Source	DF	SS	MS	F	p
A	1	18	18	2.04	0.18
B	2	48	24	2.72	0.11
A × B	2	144	72	8.15	0.006
Error	12	106	8.83		

TABLE 4.4
Rat Data: Times to Run a Maze (seconds)

	Diet		
Drug	d1	d2	d3
D1	31	36	22
	45	29	21
	46	40	18
	43	23	23
D2	82	92	30
	110	61	37
	88	49	38
	72	124	29
D3	43	44	23
	45	35	25
	63	31	24
	76	40	22
D4	45	56	30
	71	102	36
	66	71	31
	62	38	33

4.3.1. Rat Data

To provide a further example of a two-way ANOVA, the data in Table 4.4 will be used. These data arise from an experiment in which rats were randomly assigned to receive each combination of one of four drugs and one of three diets for a week, and then their time to negotiate a standard maze was recorded; four rats were used for each diet–drug combination. The ANOVA table for the data and the cell means and standard deviations are shown in Table 4.5. The observed standard deviations in this table shed some doubt on the acceptability of the homogeneity assumption, a problem we shall conveniently ignore here (but see Exercise 4.1). The F tests in Table 4.5 show that there is evidence of a difference in drug means and diet means, but no evidence of a drug × diet interaction. It appears that what is generally known as a *main effects model* provides an adequate description of these data. In other words, diet and drug (the *main effects*) act additively on time to run the maze. The results of both the Bonferonni and Scheffé multiple comparison procedures for diet and drug are given in Table 4.6. All the results are displayed graphically in Figure 4.1. Both multiple comparison procedures give very similar results for this example. The mean time to run the maze appears to differ for diets 1 and 3 and for diets 2 and 3, but not for diets 1 and 2. For the drugs, the average time is different for drug 1 and drug 2 and also for drugs 2 and 3.

When the interaction effect is nonsignificant as here, the interaction mean square becomes a further estimate of the error variance, σ^2. Consequently, there is the possibility of *pooling* the error sum of squares and interaction sum of squares to provide a possibly improved error mean square based on a larger number of degrees of freedom. In some cases the use of a pooled error mean square will provide more powerful tests of the main effects. The results of applying this procedure to the rat data are shown in Table 4.7. Here the results are very similar to those given in Table 4.5.

Although pooling is often recommended, it can be dangerous for a number of reasons. In particular, cases may arise in which the test for interaction is nonsignificant, but in which there is in fact an interaction. As a result, the pooled mean square may be larger than the original error mean square (this happens for the rats data), and the difference may be large enough for the increase in degrees of freedom to be more than offset. The net result is that the experimenter *loses* rather than gains power. Pooling is really only acceptable when there are good *a priori* reasons for assuming no interaction, and the decision to use a pooled error term is based on considerations that are independent of the observed data.

With only four observations in each cell of the table, it is clearly not possible to use, say, normal probability plots to assess the normality requirement needed within each cell by the ANOVA F tests used in Table 4.5. In general, testing the normality assumption in an ANOVA should be done *not* on the raw data, but on deviations between the observed values and the *predicted* or *fitted* values from the model applied to the data. These predicted values from a main effects model for the rat data are shown in Table 4.8; the terms are calculated from

$$\text{fitted value} = \text{grand mean} + (\text{drug mean} - \text{grand mean})$$
$$+ (\text{diet mean} - \text{grand mean}). \tag{4.1}$$

TABLE 4.5

ANOVA Table for Rat Data

Source	DF	SS	MS	F	p
Diet	2	10330.13	5165.06	23.22	<.001
Drug	3	9212.06	3070.69	13.82	<.001
Diet × Drug	6	2501.38	416.90	1.87	.11
Error	36	8007.25	222.42		

Cell Means and Standard Deviations

		Diet	
Drug	d1	d2	d3
D1			
Mean	41.25	32.00	21.00
SD	6.95	7.53	2.16
D2			
Mean	88.0	81.5	33.5
SD	16.08	33.63	4.65
D3			
Mean	56.75	37.50	23.50
SD	15.67	5.69	1.29
D4			
Mean	61.00	66.75	32.50
SD	11.28	27.09	2.65

The normality assumption may now be assessed by looking at the distribution of the differences between the observed and fitted values, the so-called *residuals*. These terms should have a normal distribution. A normal probability plot of the residuals from fitting a main efffects model to the rat data (Figure 4.2) shows that a number of the largest residuals give some cause for concern. Again, it will be left as an exercise (see Exercise 4.1) for the reader to examine this possible problem in more detail. (We shall have much more to say about residuals and their use in Chapter 6.)

4.3.2. Slimming Data

Our final example of the analysis of a two-way design involves the data shown in Table 4.9. These data arise from an investigation into types of slimming regime. In

TABLE 4.6
Multiple Comparison Test Results for Rat Data

	Estimate	Std. Error	Lower Bound	Upper Bound
Diet				
Bonferroni[a]				
d1–d2	7.31	5.27	−5.93	20.6
d1–d3	34.10	5.27	20.90	47.4[b]
d2–d3	26.80	5.27	13.60	40.1[b]
Scheffé[c]				
d1–d2	7.31	5.27	−6.15	20.8
d1–d3	34.10	5.27	20.70	47.6[b]
d2–d3	26.80	5.27	13.30	40.3[b]
Drug				
Bonferroni[d]				
D1–D2	−36.20	6.09	−53.20	−19.30[b]
D1–D3	−7.83	6.09	−24.80	9.17
D1–D4	−22.00	6.09	−39.00	−5.00[b]
D2–D3	28.40	6.09	11.40	45.40[b]
D2–D4	14.20	6.09	−2.75	31.20
D3–D4	−14.20	6.09	−31.20	2.83
Scheffé[e]				
D1–D2	−36.20	6.09	−54.1	−18.40[b]
D1–D3	−7.83	6.09	−25.7	10.00
D1–D4	−22.00	6.09	−39.9	−4.15[b]
D2–D3	28.40	6.09	10.6	46.30[b]
D2–D4	14.20	6.09	−3.6	32.10
D3–D4	−14.20	6.09	−32.0	3.69

[a]95% simultaneous confidence intervals for specified linear combinations, by the Bonferroni method; critical point: 2.511; response variable: time.

[b]Intervals that exclude zero.

[c]95% simultaneous confidence intervals for specified linear combinations, by the Scheffé method; critical point: 2.553; response variable: time.

[d]95% simultaneous confidence intervals for specified linear combinations, by the Bonferroni method; critical point: 2.792; response variable: time.

[e]95% simultaneous confidence intervals for specified linear combinations, by the Scheffé method; critical point: 2.932; response variable: time.

this case, the two factor variables are treatment with two levels, namely whether or not a woman was advised to use a slimming manual based on psychological behaviorist theory as an addition to the regular package offered by the clinic, and status, also with two levels, that is, "novice" and "experienced" slimmer. The dependent variable recorded was a measure of weight change over 3 months, with negative values indicating a decrease in weight.

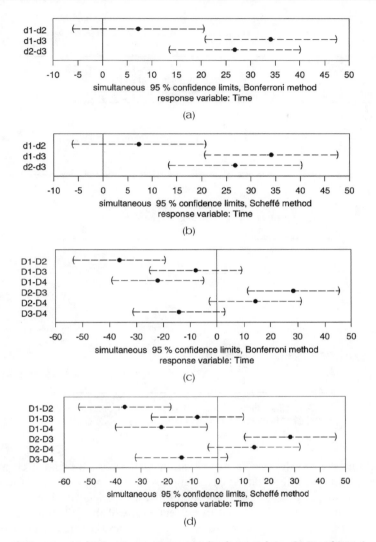

FIG. 4.1. Multiple comparison results for rat data: (a) Bonferroni tests for diets, (b) Sheffé tests for diets, (c) Bonferroni tests for drugs, (d) Scheffé tests for drugs.

The ANOVA table and the means and standard deviations for these data are shown in Table 4.10. Here the main effects of treatment and status are significant but in addition there is a significant interaction between treatment and status. The presence of a significant interaction means that some care is needed in arriving at a sensible interpretation of the results. A plot of the four cell means as shown in Figure 4.3 is generally helpful in drawing conclusions about the significant

TABLE 4.7
Pooling Error and Interaction Terms in the ANOVA of the Rat Data

Source	DF	SS	MS	F	p
Diet	2	10330.13	5165.06	20.64	<.0001
Drugs	3	9212.06	3070.69	12.27	<.0001
Pooled error	42	1058.63	250.21		

TABLE 4.8
Fitted Values for Main Effects Model on Rat Data

	Diet		
Drug	D1	D2	D3
D1	45.22	37.92	11.10
D2	81.48	74.17	47.35
D3	53.06	45.75	18.94
D4	67.23	59.92	33.10

Note. Grand mean of observations is 47.94; diet means are d1: 61.75, d2: 54.44, d3: 27.63; drug means are D1: 31.42, D2: 67.67, D3: 39.25, D4: 53.42.

interaction between the two factors. Examining the plot, we find it clear that the decrease in weight produced by giving novice slimmers access to the slimming manual is far greater than that achieved with experienced slimmers, where the reduction is negligible. Formal significance tests of the differences between experienced and novice slimmers for each treatment level could be performed—they are usually referred to as tests of *simple effects*—but in general they are unnecessary because the interpretation of a significant interaction is usually clear from a plot of cell means. (It is never wise to perform more significance tests than are absolutely essential). The significant main effects might be interpreted as indicating differences in the average weight change for novice compared with experienced slimmers, and for slimmers having access to the manual compared with those who do not. In the presence of the large interaction, however, such an interpretation is not at all helpful. The clear message from these results is as follows.

TABLE 4.9
Slimming Data

| Manual Given? | Type of Slimmer | |
	Novice	Experienced
No manual	−2.85	−2.42
	−1.98	0.00
	−2.12	−2.74
	0.00	−0.84
Manual	−4.44	0.00
	−8.11	−1.64
	−9.40	−2.40
	−3.50	−2.15

Note: Response variable is a measure of weight change over 3 months; negative values indicate a decrease in weight.

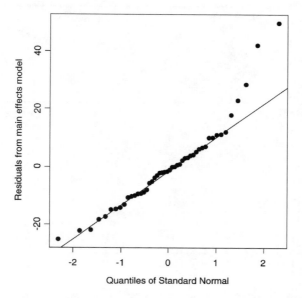

FIG. 4.2. Normal probability plot of residuals from a main effects model fitted to the rat data.

TABLE 4.10
ANOVA of Slimming Data

Analysis of Variance Table

Source	SS	DF	MS	F	p
Condition (C)	21.83	1	21.83	7.04	0.021
Status (S)	25.53	1	25.53	8.24	0.014
C × S	20.95	1	20.95	6.76	0.023
Error	37.19	12	3.10		

Cell Means and Standard Deviations

Parameter	Novice	Experienced
No manual		
Mean	−1.74	−1.50
SD	1.22	1.30
Manual		
Mean	−6.36	−1.55
SD	2.84	1.08

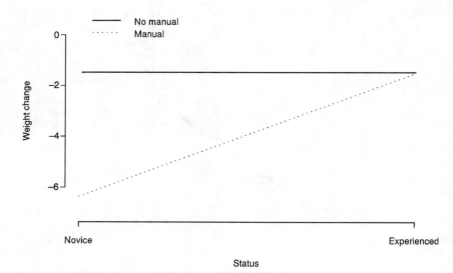

FIG. 4.3. Interaction plot for slimming data.

108

1. No manual—no difference in weight change of novice and experienced slimmers.
2. Manual—novice slimmers win!

(It might be useful to construct a confidence interval for the difference in the weight change of novice and experienced slimmers when both have access to the manual; see Exercise 4.2.)

One final point to note about the two-way design is that when there is only a single observation available in each cell, the error mean square has zero degrees of freedom. In such cases, it is generally assumed that the factors do not interact so that the interaction mean square can be used as "error" to provide F tests for the main effects.

4.4. HIGHER-ORDER FACTORIAL DESIGNS

Maxwell and Delaney (1990) and Boniface (1995) give details of an experiment in which the effects of three treatments on blood pressure were investigated. The three treatments were as follows.

1. The first is *drug*, with three levels, drug X, drug Y and drug Z;
2. The second is *biofeed*, in which biofeedback is either present or absent;
3. The third is *diet*, in which a special diet is either given or not given.

Seventy-two subjects were used in the experiment, with six being randomly allocated to each of the 12 treatment combinations. The data are shown in Table 4.11. A suitable model for a three-factor design is outlined in Display 4.2. The results of applying this model to the blood pressure data are shown in Table 4.12. The diet, biofeed, and drug main effects are all significant beyond the 5% level. None of the first-order interactions are significant, but the three way, second-order interaction of diet, drug, and biofeedback *is* significant. Just what does such an effect imply, and what are its implications for interpreting the analysis of variance results?

First, a significant second-order interaction implies that the first-order interaction between two of the variables differs in form or magnitude in the different levels of the remaining variable. Once again the simplest way of gaining insight here is to construct the appropriate graphical display. Here the interaction plots of diet and biofeedback, for each of the three drugs, will help. These plots are shown in Figure 4.4. For drug X the diet $\times$ biofeedback interaction plot indicates that diet has a negligible effect when biofeedback is supplied, but substantially reduces blood pressure when biofeedback is absent. For drug Y the situation is essentially the reverse of that for drug X. For drug Z, the blood pressure difference

TABLE 4.11
Blood Pressure Data

| Drug X | Biofeedback Present | | Biofedback Absent | | |
	Drug Y	Drug Z	Drug X	Drug Y	Drug Z
Diet Absent					
170	186	180	173	189	202
175	194	187	194	194	228
165	201	199	197	217	190
180	215	170	190	206	206
160	219	204	176	199	224
158	209	194	198	195	204
Diet Present					
161	164	162	164	171	205
173	166	184	190	173	199
157	159	183	169	196	170
152	182	156	164	199	160
181	187	180	176	180	179
190	174	173	175	203	179

Display 4.2
Model for a Three-Factor Design with Factors A, B, and C

- In general terms the model is

    Observed response = mean + factor A effect + factor B effect
    + factor C effect + AB interaction
    + AC interaction + BC interaction
    + ABC interaction + error.

- More specifically, let y_{ijkl} represent the lth observation in the kth level of factor C (with c levels), the jth level of factor B (with b levels), and the ith level of factor A (with a levels). Assume that there are n subjects per cell, and that the factor levels are of specific interest so that A, B, and C have fixed effects.
- The linear model for the observations in this case is

$$y_{ijkl} = \mu + \alpha_i + \beta_j + \gamma_k + \delta_{ij} + \tau_{ik} + \omega_{jk} + \theta_{ijk} + \epsilon_{ijkl}.$$

Here α_i, β_j, and γ_k represent main effects of A, B, and C, respectively; δ_{ij}, τ_{ik}, and ω_{jk} represent *first-order* interactions, AB, AC, and BC; θ_{ijkl} represents *second-order* interaction, ABC; and ϵ_{ijkl} are random error terms assumed to be normally distributed with zero mean and variance σ^2. (Once again the parameters have to be constrained in some way to make the model acceptable); see Maxwell and Delaney, 1990, for details.)

(Continued)

Display 4.2

(Continued)

- The null hypotheses here are that there are no main effects of A, B, or C, no first-order interactions, AB, AC, or BC, and no second-order interaction, ABC.
- In terms of the parameters in the model, these hypotheses are

$$H_0^{(A)} : \alpha_1 = \alpha_2 = \cdots = \alpha_a = 0,$$

$$H_0^{(B)} : \beta_1 = \beta_2 = \cdots = \beta_b = 0,$$

$$H_0^{(C)} : \gamma_1 = \gamma_2 = \cdots = \gamma_c = 0,$$

$$H_0^{(AB)} : \delta_{11} = \delta_{12} = \cdots = \delta_{ab} = 0,$$

$$H_0^{(AC)} : \tau_{11} = \tau_{12} = \cdots = \tau_{ac} = 0,$$

$$H_0^{(BC)} : \omega_{11} = \omega_{12} = \cdots = \omega_{bc} = 0,$$

$$H_0^{(ABC)} : \theta_{111} = \theta_{112} = \cdots = \theta_{abc} = 0.$$

- The analysis of variance table for the design is

Source	SS	DF	MS	F
A	ASS	$a - 1$	$\frac{ASS}{(a-1)}$	MSA/error MS
B	BSS	$b - 1$	$\frac{BSS}{(b-1)}$	MSB/error MS
C	CSS	$c - 1$	$\frac{CSS}{(c-1)}$	MSC/error MS
A × B	ABSS	$(a-1)(b-1)$	$\frac{ABSS}{(a-1)(b-1)}$	MSAB/error MS
A × C	ACSS	$(a-1)(c-1)$	$\frac{ACSS}{(a-1)(c-1)}$	MSAC/error MS
B × C	BCSS	$(b-1)(c-1)$	$\frac{BCSS}{(b-1)(c-1)}$	BCMS/error MS
A × B × C	ABCSS	$(a-1)(b-1)(c-1)$	$\frac{ABCSS}{(a-1)(b-1)(c-1)}$	MSABC/error MS
Error (within cell)	WCSS	$abc(n-1)$	$\frac{WCSS}{abc(n-1)}$	

when the diet is given and when it is not is approximately equal for both levels of biofeedback.

The implication of a significant second-order interaction is that there is little point in drawing conclusions about either the nonsignificant first-order interaction or the significant main effects. The effect of drug, for example, is not consistent for all combinations of diet and biofeedback. It would, therefore, be misleading to conclude on the basis of the significant main effect anything about the specific effects of these three drugs on blood pressure. The interpretation of the data might become simpler by carrying out separate two-way analyses of variance within each drug (see Exercise 4.3).

Clearly, factorial designs will become increasingly complex as the number of factors is increased. A further problem is that the number of subjects required for

TABLE 4.12
ANOVA for Blood Pressure Data

Source	SS	DF	MS	F	p
Diet	5202	1	5202.0	33.20	<.001
Biofeed	2048	1	2048.0	13.07	<.001
Drug	3675	2	1837.5	11.73	<.001
Diet × biofeed	32	1	32.0	0.20	.66
Diet × drug	903	2	451.5	2.88	.06
Biofeed × drug	259	2	129.5	0.83	.44
Diet × biofeed × drug	1075	2	537.5	3.42	.04
Error	9400	60	156.67		

a complete factorial design quickly becomes prohibitively large. Consequently, alternative designs have to be considered that are more economical in terms of subjects. Perhaps the most common of these is the *latin square* (described in detail by Maxwell and Delaney, 1990). In such a design, economy in number of subjects required is achieved by assuming *a priori* that there are no interactions between factors.

4.5. RANDOM EFFECTS AND FIXED EFFECTS MODELS

In the summary of Chapter 3, a passing reference was made to *random* and *fixed* effects factors. It is now time to consider these terms in a little more detail, beginning with some formal definitions.

1. A factor is random if its levels consist of a random sample of levels from a population of all possible levels.
2. A factor is fixed if its levels are selected by a nonrandom process or its levels consist of the entire population of possible levels.

Three basic types of models can be constructed, depending on what types of factors are involved.

1. A model is called a *fixed effects model* if all the factors in the design have fixed effects.
2. A model is called a *random effects model* if all the factors in the design have random effects.

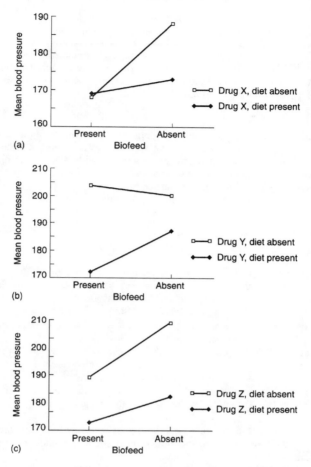

FIG. 4.4. Interaction plots for blood pressure data.

3. A model is called a *mixed effects model* if some of the factors in the design have fixed effects and some have random effects.

No doubt readers will not find this wildly exciting until some explanation is at hand as to why it is necessary to differentiate between factors in this way. In fact, the philosophy behind random effects models is quite different from that behind the use of fixed effects models, both in the sampling scheme used and in the parameters of interest. The difference can be described relatively simply by using a two-way design; a suitable model for such a design when both factors are regarded as random is shown in Display 4.3. The important point to note about this model, when compared with that for fixed effect factors (see Display 4.1), is that the terms α_i,

Display 4.3

Random Effects Model for a Two-Way Design with Factors A and B

- Let y_{ijk} represent the kth observation in the jth level of factor B (with b levels randomly sampled from some population of possible levels) and the ith level of factor A (with a levels again randomly sampled from a population of possible levels).
- The linear model for the observations is

$$y_{ijk} = \mu + \alpha_i + \beta_j + \gamma_{ij} + \epsilon_{ijk},$$

where now the α_i are assumed to be random variables from a normal distribution with mean zero and variance σ_α^2. Similarly, β_j and γ_{ij} are assumed to be random variables from normal distributions and zero means with variances σ_β^2 and σ_γ^2, respectively. As usual the ϵ_{ijk} are error terms randomly sampled from a normal distribution with mean zero and variance σ^2.

- The hypotheses of interest in the random effects model are

$$H_0^{(\alpha)} : \sigma_\alpha^2 = 0$$
$$H_0^{(\beta)} : \sigma_\beta^2 = 0$$
$$H_0^{(\gamma)} : \sigma_\gamma^2 = 0.$$

- The calculation of the sums of squares and mean squares remains the same as for the fixed effects model. Now, however, the various mean squares provide estimators of combinations of the variances of the parameters: (a) A mean square, estimator of $\sigma^2 + n\sigma_\gamma^2 + nb\sigma_\alpha^2$; (b) B mean square, estimator of $\sigma^2 + n\sigma_\gamma^2 + na\sigma_\beta^2$; (c) AB mean square, estimator of $\sigma^2 + n\sigma_\gamma^2$; and (d) error mean square, as usual, estimator of σ^2.
- If $H_0^{(\gamma)}$ is true, then the AB mean square and the error mean square both estimate σ^2. Consequently, the ratio of the AB mean square and the error mean square can be tested as an F statistic to provide a test of the hypothesis.
- If $H_0^{(\alpha)}$ is true, the A mean square and the AB mean square both estimate $\sigma^2 + n\sigma_\gamma^2$, so the ratio of the A mean square and the AB mean square provides an F test of the hypothesis.
- If $H_0^{(\beta)}$ is true, the B mean square and the AB mean square both estimate $\sigma^2 + \sigma_\gamma^2$, so the ratio of the B mean square and the AB mean square provides an F test of the hypothesis
- Estimators of the variances σ_α^2, σ_β^2, and σ_γ^2 can be obtained from

$$\hat{\sigma}_\alpha^2 = \frac{\text{MSA} - \text{MSAB}}{nb},$$

$$\hat{\sigma}_\beta^2 = \frac{\text{MSB} - \text{MSAB}}{na},$$

$$\hat{\sigma}_\gamma^2 = \frac{\text{MSAB} - \text{error MS}}{n}.$$

β_j, and γ_{ij} are now assumed to be *random variables* with particular distributions. The parameters of interest are the *variances* of these distributions, and they can be estimated as shown in Display 4.3. The terms in the analysis of variance table are calculated as for the fixed effects model described in Display 4.1, but the F tests for assessing whether the main effect variances are zero now involve the interaction mean square rather than the error mean square. However, the test for whether the

TABLE 4.13
ANOVA of Rat Data by Using a Random Effects Model

Source	DF	SS	MS	F	p
Diets	2	10330.13	5165.06	12.39	<.0001
Drugs	3	9212.06	3070.69	7.37	<.001
Diets × drugs	6	2501.38	416.90	1.87	.11
Error	36	8007.25	222.42		

Note.

$$\hat{\sigma}_\alpha^2 = \frac{5165.06 - 416.90}{16} = 296.76,$$

$$\hat{\sigma}_\beta^2 = \frac{3070.69 - 416.90}{12} = 221.15,$$

$$\hat{\sigma}_\gamma^2 = \frac{416.90 - 222.42}{4} = 48.62.$$

variance of the distribution of the interaction parameters is zero is the same as that used for testing the hypothesis of no interaction in the fixed effects model.

To illustrate the application of a random effects model, the data in Table 4.3 will again be used, making the unrealistic assumption that both the particular diets and the particular drugs used are a random sample from a large population of possible diets and drugs. The resulting analysis is shown in Table 4.13. The F tests indicate that the variances of both the drug and diet parameter distributions are not zero; however, the hypothesis that $\sigma_\gamma^2 = 0$ is not rejected.

The main differences between using fixed effects and random effects models are as follows.

1. Because the levels of a random effects factor have been chosen randomly, the experimenter is not interested in the means of the particular levels observed. In particular, planned or *post hoc* comparisons are no longer relevant.
2. In a random effects model, interest lies in the estimation and testing of variance parameters.
3. An advantage of a random effects model is that it allows the experimenter to generalize beyond the levels of factors actually observed.

It must be emphasized that generalizations to a population of levels from the tests of significance for a random effects factor are warranted *only* when the levels of the factor are actually selected at random from a population of levels. It seems unlikely that such a situation will hold in most psychological experiments that use factorial designs, and, consequently, fixed effects models are those most generally used except in particular situations, as we shall see in Chapters 5 and 7.

4.6. FACTORIAL DESIGNS WITH UNEQUAL NUMBERS OF OBSERVATIONS IN EACH CELL

More observant readers will have noticed that all the examples discussed in the previous sections of this chapter have had the same number of observations in each cell—they have been *balanced designs*. In most experimental situations, equality of cell size is the aim, although even in well-designed experiments, things can, of course, go wrong. A researcher may be left with an experiment having unequal numbers of observations in each cell as a result of the death of an experimental animal, for example, or because of the failure of a subject to attend a planned experimental session. In this way, an aimed-for balanced design can become *unbalanced*, although in an experimental situation the imbalance in cell sizes is likely to be small.

In observational studies, however, far larger inequalities in cell size may arise. Two examples will serve to illustrate this point.

First, the data shown in Table 4.14 come from a sociological study of Australian Aboriginal and White children reported by Quine (1975). In the study, children of both sexes, from four age groups (final grade in primary schools and first, second, and third form in secondary school), and from two cultural groups were used. The children in each age group were classified as slow or average learners. The basic design of the study is then an unequally replicated $4 \times 2 \times 2 \times 2$ factorial. The response variable of interest was the number of days absent from school during the school year. Children who had suffered a serious illness during the year were excluded.

Second, the data for the second example are shown in Table 4.15. These data arise from an investigation into the effect of a mother's postnatal depression on child development. Mothers who give birth to their first-born child in a major teaching hospital in London were divided into two groups, depressed and not depressed, on the basis of their mental state 3 months after the birth. The children's fathers were also divided into two groups, namely those who had a previous history of psychiatric illness and those who did not. The dependent variable to be considered first is the child's IQ at age 4. (The other variables in Table 4.15 will be used later in this chapter.)

So unbalanced factorial designs do occur. But does it matter? Why do such designs merit special consideration? Why is the usual ANOVA approach used earlier in this chapter not appropriate? The answers to all these questions lie in recognizing that in unbalanced factorial designs the factors forming the design

TABLE 4.14
Study of Australian Aboriginal and White Children

Cell	Origin	Sex	Grade	Type	Days
1	A	M	F0	SL	2, 11, 14
2	A	M	F0	AL	5, 5, 13, 20, 22
3	A	M	F1	SL	6, 6, 15
4	A	M	F1	AL	7, 14
5	A	M	F2	SL	6, 32, 53, 57
6	A	M	F2	AL	14, 16, 16, 17, 40, 43, 46
7	A	M	F3	SL	12, 15
8	A	M	F3	AL	8, 23, 23, 28, 34, 36, 38
9	A	F	F0	SL	3
10	A	F	F0	AL	5, 11, 24, 45
11	A	F	F1	SL	5, 6, 6, 9, 13, 23, 25, 32, 53, 54
12	A	F	F1	AL	5, 5, 11, 17, 19
13	A	F	F2	SL	8, 13, 14, 20, 47, 48, 60, 81
14	A	F	F2	AL	2
15	A	F	F3	SL	5, 9, 7
16	A	F	F3	AL	0, 2, 3, 5, 10, 14, 21, 36, 40
17	N	M	F0	SL	6, 17, 67
18	N	M	F0	AL	0, 0, 2, 7, 11, 12
19	N	M	F1	SL	0, 0, 5, 5, 5, 11, 17
20	N	M	F1	AL	3, 3
21	N	M	F2	SL	22, 30, 36
22	N	M	F2	AL	8, 0, 1, 5, 7, 16, 27
23	N	M	F3	SL	12, 15
24	N	M	F3	AL	0, 30, 10, 14, 27, 41, 69
25	N	F	F0	SL	25
26	N	F	F0	AL	10, 11, 20, 33
27	N	F	F1	SL	5, 7, 0, 1, 5, 5, 5, 5, 7, 11, 15
28	N	F	F1	AL	5, 14, 6, 6, 7, 28
29	N	F	F2	SL	0, 5, 14, 2, 2, 3, 8, 10, 12
30	N	F	F2	AL	1
31	N	F	F3	SL	8
32	N	F	F3	AL	1, 9, 22, 3, 3, 5, 15, 18, 22, 37

Note. Response variable is days away from school. A, aboriginal; N, nonaboriginal; F, female; M, male; F0, primary; F1, first form; F2, second form; F3, third form; SL, slow learner; FL, fast learner.

TABLE 4.15
Data Obtained in a Study of the Effect of Postnatal Depression of the Mother on a
Child's Cognitive Development

Mother	Depression	Sex of Child	IQ	PS	VS	QS	Husband's Psychiatric History
1	ND	Boy	103	56	50	42	No history
2	ND	Boy	124	65	64	61	No history
3	ND	Boy	124	67	61	63	No history
4	ND	Girl	104	52	55	44	No history
5	D	Girl	96	46	52	46	No history
6	ND	Boy	92	46	43	43	No history
7	ND	Girl	124	58	68	65	No history
8	ND	Girl	99	55	46	50	No history
9	ND	Boy	92	46	41	55	No history
10	ND	Boy	116	61	58	51	No history
11	ND	Boy	99	58	45	50	No history
12	D	Girl	22	22	50	22	History
13	D	Girl	81	47	38	41	History
14	ND	Girl	117	68	53	63	No history
15	D	Girl	100	54	47	53	History
16	ND	Girl	89	48	41	46	No history
17	ND	Girl	125	64	66	67	No history
18	ND	Boy	127	64	68	71	History
19	ND	Girl	112	50	57	64	No history
20	ND	Girl	48	23	20	25	No history
21	ND	Boy	139	68	75	64	No history
22	ND	Girl	118	58	64	61	No history
23	D	Girl	107	45	58	58	History
24	ND	Boy	106	46	57	53	No history
25	D	Boy	129	63	70	67	No history
26	ND	Boy	117	43	71	63	No history
27	ND	Boy	123	64	64	61	History
28	ND	Girl	118	64	56	60	No history
29	D	Boy	84	47	37	43	History
30	ND	Boy	117	66	64	61	No history
31	D	Girl	102	48	52	52	No history
32	ND	Boy	141	66	76	69	No history
33	ND	Boy	124	60	77	58	No history
34	ND	Boy	110	50	61	52	No history
35	ND	Boy	98	54	47	48	No history
36	ND	Boy	109	64	48	50	No history
37	ND	Boy	120	71	53	63	No history
38	ND	Boy	127	67	68	59	No history
39	ND	Girl	103	55	52	48	No history

(Continued)

TABLE 4.15
(Continued)

Mother	Depression	Sex of Child	IQ	PS	VS	QS	Husband's Psychiatric History
40	ND	Boy	118	65	57	61	No history
41	ND	Boy	117	60	64	48	No history
42	ND	Boy	115	52	67	55	No history
43	ND	Boy	119	63	62	58	No history
44	ND	Boy	117	63	56	59	No history
45	ND	Boy	92	45	42	55	No history
46	ND	Girl	101	53	48	52	No history
47	ND	Boy	119	59	65	59	No history
48	ND	Boy	144	78	67	65	No history
49	ND	Boy	119	66	54	62	No history
50	ND	Boy	127	67	63	67	No history
51	ND	Girl	113	61	54	58	No history
52	ND	Boy	127	60	59	68	No history
53	ND	Boy	103	57	50	48	No history
54	ND	Girl	128	70	65	60	No history
55	ND	Girl	86	45	45	28	No history
56	ND	Boy	112	62	51	60	No history
57	ND	Girl	115	59	58	59	No history
58	ND	Girl	117	48	68	62	No history
59	ND	Girl	99	58	48	41	No history
60	ND	Girl	110	54	61	47	No history
61	D	Girl	139	70	72	66	No history
62	ND	Girl	117	64	57	51	No history
63	ND	Boy	96	52	45	47	No history
64	D	Boy	111	54	60	50	No history
65	ND	Boy	118	58	62	62	No history
66	ND	Boy	126	66	64	67	No history
67	ND	Girl	126	69	66	55	No history
68	ND	Girl	89	49	36	36	No history
69	ND	Boy	102	56	49	50	No history
70	ND	Boy	134	74	68	59	No history
71	ND	Boy	93	47	46	45	No history
72	ND	Boy	115	55	61	60	No history
73	D	Girl	99	50	48	54	History
74	ND	Girl	99	58	44	47	No history
75	ND	Girl	122	55	64	74	No history
76	ND	Girl	106	49	54	59	History
77	ND	Girl	124	66	58	68	No history
78	ND	Boy	100	43	56	48	No history
79	ND	Boy	114	61	56	55	No history

(Continued)

TABLE 4.15

(Continued)

Mother	Depression	Sex of Child	IQ	PS	VS	QS	Husband's Psychiatric History
80	ND	Boy	121	64	66	50	No history
81	ND	Boy	119	63	61	60	No history
82	ND	Boy	108	48	60	50	No history
83	ND	Boy	110	66	56	48	No history
84	ND	Boy	127	62	71	61	No history
85	ND	Girl	118	66	56	58	No history
86	ND	Girl	107	53	54	52	No history
87	D	Girl	123	62	68	59	No history
88	D	Girl	102	48	52	58	No history
89	ND	Girl	110	65	47	55	History
90	ND	Boy	114	64	56	47	No history
91	ND	Girl	118	58	63	52	No history
92	D	Boy	101	44	56	47	No history
93	D	Boy	121	64	62	62	No history
94	ND	Girl	114	50	65	51	No history

Note. ND, not depressed; D, depressed; PS, perceptual score; VS, verbal score; QS, quantitative score.

are no longer independent as they are in a balanced design. For example, the 2×2 table of counts of children of depressed (D) and nondepressed (ND) mothers and "well" and "ill" fathers for the data in table in 4.15 is

	Mother		
Father	ND	D	Total
Well	75(70)	9(14)	84
Ill	4(8)	6(2)	10
Total	79	15	94

The numbers in parenthesis show that the expected number of observations (rounded to the nearest whole number) under the hypothesis that the mother's state is independent of the father's state. Clearly there are more depressed mothers whose husbands are psychiatrically ill than expected under independence. (Review the coverage of the chi-square test given in your introductory course and in Chapter 9 for details of how expected values are calculated.)

The results of this dependence of the factor variables is that it is now no longer possible to partition the total variation in the response variable into *nonoverlapping* or *orthogonal* sums of squares representing factor main effects and interactions. For example, in an unbalanced two-way design, there is a proportion of the variance of the response variable that can be attributed to (*explained by*) either factor A or factor B. A consequence is that A and B together explain less of the variation of the dependent variable than the sum of which each explains alone. The result is the sum of squares that can be attributed to a factor depends on which other factors have already been allocated a sum of squares; in other words, the sums of squares of factors depend on the order in which they are considered. (This point will be taken up in more detail in the discussion of regression analysis in Chapter 6.)

The dependence between the factor variables in an unbalanced factorial design, and the consequent lack of uniqueness in partitioning the variation in the response variable, has led to a great deal of confusion about what is the most appropriate analysis of such designs. The issues are not straightforward and even statisticians (yes, even statisticians!) are not wholly agreed on the most suitable method of analysis for all situations, as is witnessed by the discussion following the papers of Nelder (1977) and Aitkin (1978).

Basically the discussion over the analysis of unbalanced factorial designs has involved the question of whether what are called *sequential* or *Type I* sums of squares, or *unique* or *Type III* sums of squares, should be used. In terms of a two-way design with factors A and B, these sums of squares are as follows.

Type I Sum of Squares
(Sequential Sums of Squares)

Here the sums of squares come from fitting the terms in the model sequentially, giving, for example,

Sum of squares for A,
Sum of squares for B after A (usually denoted as B|A),
Sum of squares for AB after A and B (AB|A, B).

To establish a suitable model for the data (see later), we would need to calculate both the sums of squares above, and these with the order of A and B reversed, that is,

Sum of squares for B,
Sum of squares for A|B,
Sum of squares for AB|B, A.

(The interaction sum of squares would be identical in both cases.)

Type III Sum of Squares
(Unique Sums of Squares)

Here the sum of squares come from fitting each term in the model, *after* all other terms; these sum of squares are said to be unique for the particular term. Thus in our two-way design we would have

Sum of squares for A = A|B, AB,
Sum of squares for B = B|A, AB,
Sum of squares for AB = AB|A, B.

(Again, the interaction sum of squares will be the same as in the sequential approach.)

In a balanced design, Type I and Type III sums of squares are identical. How the respective sums of squares are calculated will be come clear in Chapter 6, but the question here is, Which type should be used? Well, if you take as your authority Professor David Howell in his best-selling text book, *Statistical Methods for Psychology*, then there is no doubt—you should always use Type III sums of squares and never Type I. And in Maxwell and Delaney's *Designing Experiments and Analysing Data*, the same recommendation appears, although somewhat less strongly. However, in the papers by Nelder and Aitkin referred to earlier, adjusting main effects for interaction as used in Type III sums of squares is severely criticized both on theoretical and pragmatic grounds. The arguments are relatively subtle, but in essence they go something like this.

1. When models are fit to data, the principle of parsimony is of critical importance. In choosing among possible models, we do not adopt complex models for which there is no empirical evidence.
2. So if there is no convincing evidence of an AB interaction, we do not retain the term in the model. Thus additivity of A and B is assumed unless there is convincing evidence to the contrary.
3. The issue does not arise as clearly in the balanced case, for there the sum of squares for A, say, is independent of whether interaction is assumed or not. Thus in deciding on possible models for the data, we do not include the interaction term unless it has been shown to be necessary, in which case tests on main effects involved in the interaction are not carried out (or if carried out, not interpreted; see biofeedback example).
4. Thus the argument proceeds that Type III sum of squares for A in which it is adjusted for AB as well as B makes no sense.
5. First, if the interaction term is necessary in the model, then the experimenter will usually wish to consider simple effects of A at each level of B separately. A test of the hypothesis of no A main effect would not usually be carried out if the AB interaction is significant.

TABLE 4.16
ANOVA of Postnatal Depression Data

Source	DF	SS	MS	F	p
Order 1					
Mother	1	1711.40	1711.40	6.82	.01
Husband	1	1323.29	1323.29	5.27	.02
Mother × Husband	1	2332.21	2332.21	9.29	.003
Error	90	22583.58	250.93		
Order 2					
Husband	1	2526.44	2526.44	10.07	0.002
Mother	1	508.26	508.25	2.025	0.16
Husband × Mother	1	2332.21	2332.21	9.29	0.003
Error	90	22583.58	250.93		

6. If the AB interaction is not significant, then adjusting for it is of no interest and causes a substantial loss of power in testing the A and B main effects.

The arguments of Nelder and Aitkin against the use of Type III sums of squares are persuasive and powerful. Their recommendation to use Type I sums of squares, perhaps considering effects in a number of orders, as the most suitable way in which to identify a suitable model for a data set is also convincing and strongly endorsed by this author.

Let us now consider the analysis of the two unbalanced data sets introduced earlier, beginning with the one involving postnatal depression and child's IQ. We shall consider only the factors of mother's mental state and husband's mental state here. The results of two analyses using Type I sums of squares are shown in Table 4.16. Here it is the interaction term that is the key, and a plot of average IQ in the four cells of the design is shown in Figure 4.5. It appears that the effect on child's IQ of a father with a psychiatric history is small if the mother is not depressed, but when the mother is depressed, there is a substantially lower IQ. (Here we have not graphed the data in any way prior to our analysis—this may have been a mistake! See Exercise 4.5.)

The Australian data given in Table 4.14 are complex, and here we shall only scratch the surface of possible analyses. (A very detailed discussion of the analysis of these data is given in Aitkin, 1978.) An analysis of variance table for a particular order of effects is given in Table 4.17. The significant origin × sex × type interaction is the first term that requires explanation; this task is left to readers (see Exercise 4.6).

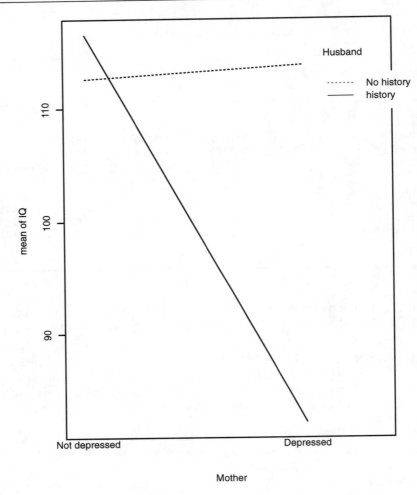

FIG. 4.5. Interaction plot for postnatal depression data.

In Table 4.18 is the default SPSS analysis of variance table for these data, which, inappropriately for the reasons given above, gives Type III sums of squares. Look, for example, at the origin line in this table and compare it with that in Table 4.17. The difference is large; adjusting the origin effect for the large number of mostly nonsignificant interactions has dramatically reduced the origin sum of squares and the associated F test value. Only the values in the lines corresponding to the origin $\times$ sex $\times$ grade $\times$ type effect and the error term are common to Tables 4.17 and 4.18.

TABLE 4.17
Analysis of Data on Australian Children

Source	DF	SS	MS	F	p
		Order 1			
Origin	1	2637.69	2637.69	13.68	<.001
Sex	1	342.09	342.09	1.77	.19
Grade	3	1827.77	609.26	3.16	.03
Type	1	129.57	129.57	0.67	.41
Origin × Sex	1	144.44	144.44	0.75	.39
Origin × Grade	3	3152.79	1050.93	5.45	.002
Sex × Grade	3	1998.91	666.30	3.46	.019
Origin × Type	1	3.12	3.12	0.02	.90
Sex × Type	1	76.49	76.49	0.40	.53
Grade × Type	3	2171.45	723.82	3.75	.013
Origin × Sex × Grade	3	221.54	73.85	0.38	.77
Origin × Sex × Type	1	824.03	824.03	4.27	.04
Origin × Grade × Type	3	895.69	298.56	1.55	.21
Sex × Grade × Type	3	378.47	126.16	0.65	.58
Origin × Sex × Grade × Type	3	349.60	116.53	0.60	.61
Error	122	23527.22	192.85		

TABLE 4.18
Default ANOVA Table Given by SPSS for Data on Australian School Children

Source	DF	SS	MS	F	p
Origin	1	228.72	228.72	1.19	.28
Sex	1	349.53	349.53	1.762	.18
Grade	3	1019.43	339.81	1.762	.16
Type	1	177.104	177.04	0.92	.34
Origin × Sex	1	3.82	3.82	0.02	.89
Origin × Grade	3	1361.06	453.69	2.35	.08
Origin × Type	1	53.21	53.21	0.28	.60
Sex × Type	1	0.86	0.86	0.00	.95
Grade × Type	2	2038.98	679.66	3.52	.02
Sex × Grade	3	1612.12	537.37	2.79	.04
Origin × Sex × Grade	3	34.87	11.62	0.06	.98
Origin × Grade × Type	3	980.47	326.83	1.72	.19
Sex × Grade × Type	3	412.46	137.49	0.71	.55
Origin × Sex × Grade × Type	3	349.60	116.53	0.60	.61
Error	122	23527.22	192.85		

4.7. ANALYSIS OF COVARIANCE IN FACTORIAL DESIGNS

Analysis of covariance in the context of a one-way design was introduced in Chapter 3. The technique can be generalized to factorial designs, and in this section an example of its use in a 3×3 two-way design will be illustrated by using data from Howell (1992). The data are given in Table 4.19 and arise from an experiment in which subjects performed either a pattern recognition task, a cognitive task,

TABLE 4.19
Smoking and Performance Data

Pattern Recognition

NS

Errors	9	8	12	10	7	10	9	11	8	10	8	10	8	11	10
Distract	107	133	123	94	83	86	112	117	130	111	102	120	118	134	97

DS

Errors	12	7	14	4	8	11	16	17	5	6	9	6	6	7	16
Distract	101	75	138	94	138	127	126	124	100	103	120	91	138	88	118

AS

Errors	8	8	9	1	9	7	16	19	1	1	22	12	18	8	10
Distract	64	135	130	106	123	117	124	141	95	98	95	103	134	119	123

Cognitive Task

NS

Errors	27	34	19	20	56	35	23	37	4	30	4	42	34	19	49
Distract	126	154	113	87	125	130	103	139	85	131	98	107	107	96	143

DS

Errors	48	29	34	6	18	63	9	54	28	71	60	54	51	25	49
Distract	113	100	114	74	76	162	80	118	99	146	132	135	111	106	96

AS

Errors	34	65	55	33	42	54	21	44	61	38	75	61	51	32	47
Distract	108	191	112	98	128	145	76	107	128	142	144	131	110	132	

Driving Simulation

NS

Errors	15	2	2	14	5	0	16	14	9	17	15	9	3	15	13
Distract	110	96	112	114	137	125	168	102	109	111	137	106	117	101	116

DS

Errors	7	0	6	0	12	17	1	11	4	4	3	5	16	5	11
Distract	93	102	108	100	123	131	99	116	81	103	78	103	139	101	102

AS

Errors	3	2	0	0	6	2	0	6	4	1	0	0	6	2	3
Distract	130	83	91	92	109	106	99	109	136	102	119	84	68	67	114

AS, active smoking; DS, delayed smoking; NS, nonsmoking.

or a driving simulation task, under three smoking conditions. These were *active smoking*, smoked during or just before the task; *delayed smoking*, not smoked for 3 hours; and *nonsmoking*, which consisted of nonsmokers. The dependent variable was the number of errors on the task. Each subject was also given a distractibility score—higher scores indicate a greater possibility of being distracted.

In his analysis of these data, Howell gives the ANOVA table shown in Table 4.20. This table arises from using SPSS and selecting UNIQUE sum of squares (the default). The arguments against such sums of squares are the same here as in the previous section, and a preferable analysis of these data involves Type I sums of squares with perhaps the covariate ordered first and the group × task interaction ordered last. The results of this approach are shown in Table 4.21. There is strong evidence of a group × task interaction, which again we leave as an exercise for the reader to investigate in more detail. (It is this conflict between Type I and Type III sums of squares that led to the different analyses of covariance results given by SPSS and S-PLUS on the WISC data reported in the previous chapter.)

TABLE 4.20
ANCOVA of Pattern Recognition Experiment by Using Type III Sums of Squares

Source	DF	SS	MS	F	p
Covariate (distract)	1	4644.88	464.88	64.93	.000
Group	2	563.26	281.63	3.94	.022
Task	2	23870.49	11935.24	166.84	.000
Group × Task	4	1626.51	406.63	5.68	.000
Error	125	8942.32	71.54		

TABLE 4.21
ANCOVA of Pattern Recognition Data by Using Type I Sums of Squares with the Covariate Ordered First

Source	DF	SS	MS	F	p
Distract	1	10450.43	10450.43	146.08	<.001
Task	2	23726.78	11863.39	165.83	<.001
Group	2	585.88	292.94	4.10	.02
Group × Task	4	1626.51	406.63	5.68	<.001
Error	125	8942.32	71.54		

4.8. MULTIVARIATE ANALYSIS
OF VARIANCE FOR FACTORIAL DESIGNS

In the study of the effect of mother's postnatal depression on child development, a number of cognitive variables in addition to IQ were recorded for each child at age 4 and are given in Table 4.14. Separate analyses of variance could be carried out on these variables (using Type I sums of squares), but in this case the investigator may genuinely be more interested in answering questions about the *set* of three variables. Such an approach might be sensible if the variables were to be regarded as indicators of some more fundamental concept not directly measurable—a so called *latent variable*. The required analysis is now a multivariate analysis of variance, as described in the previous chapter. The results are shown in Table 4.22. (Here we have included sex of child as an extra factor.) Note that because the design is unbalanced, order of effects is important. Also note that because each term in the ANOVA table has only a single degree of freedom, the four possible multivariate test statistics defined in the glossary all lead to the same approximate F values and the same p values. Consequently, only the value of the Pillai test statistic is given in Table 4.22. For this set of cognitive variables only the mother's mental state is significant. The means for the three variables for depressed and nondperessed mothers are as follows.

Mother	PS	VS	QS
D	50.9	54.8	51.9
ND	53.3	57.1	55.3

The children of depressed mothers have lower average scores on each of the three cognitive variables.

TABLE 4.22
MANOVA of Cognitive Variables from a Postnatal Depression Study

Source	Pillai Statistic	Approx. F	DF1	DF2	p
Mother	0.09	2.83	3	84	.04
Husband	0.05	1.45	3	84	.23
Sex	0.05	1.42	3	84	.24
Mother × Husband	0.07	2.25	3	84	.09
Mother × Sex	0.01	0.31	3	84	.82
Husband × Sex	0.002	0.07	3	84	.98
Mother × Husband × Sex	0.04	1.11	3	84	.35

4.9.   SUMMARY

1. Factorial designs allow interactions between factors to be studied.
2. The factors in factorial designs can be assumed to have random or fixed effects. The latter are generally more applicable to psychological experiments, but random effects will be of more importance in the next chapter and in Chapter 7.
3. Planned comparison and multiple comparison tests can be applied to factorial designs in a similar way to that described for a one-way design in Chapter 3.
4. Unbalanced factorial designs require care in their analyses. It is important to be aware of what procedure for calculating sums of squares is being employed by whatever statistical software is to be used.

COMPUTER HINTS

Essentially, the same hints apply as those given in Chapter 3. The most important thing to keep a look out for is the differences in how the various packages deal with unbalanced factorial designs. Those such as SPSS that give Type III sums of squares as their default should be used with caution, and then only after specifically requesting Type I sums of squares and then considering main effects before interactions, first-order interactions before second-order interactions, and so on. When S-PLUS is used for the analysis of variance of unbalanced designs, only Type I sums of squares are provided.

EXERCISES

4.1. Reexamine and, if you consider it necessary, reanalyze the rat data in the light of the differing cell standard deviations and the nonnormality of the residuals as reported in the text.

4.2. Using the slimming data in Table 4.9, construct the appropriate 95% confidence interval for the difference in weight change between novice and experienced slimmers when members of both groups have access to a slimming manual.

4.3. In the drug, biofeed, and diet example, carry out separate two-way analyses of variance of the data corresponding to each drug. What conclusions do you reach from your analysis?

4.4. In the postnatal depression data, separately analyze each of the three variables used in the MANOVA in Section 4.7. Comment on the results.

4.5. Produce box plots of the IQ score in the postnatal depression data for the four cells in the 2×2 design with factors of father's psychiatric history and

Display 4.4
ANCOVA Model for a Two-Way Design with Factors A and B

- In general terms the model is

  observed response = mean + factor A effect
                        + factor B effect + AB interaction effect
                        + covariate effect + error.

- More specifically, the analysis of covariance model for a two-way design is

$$y_{ijk} = \mu + \alpha_i + \theta_j + \gamma_{ij} + \beta(x_{ijk} - \bar{x}) + \epsilon_{ijk},$$

  where y_{ijk} is the kth observed value of the response variable in the ijth cell of the design and x_{ijk} is the corresponding value of the covariate, which has grand mean $\bar{x}$. The regression coefficient linking the response and covariate variable is β. The other terms correspond to those described in Display 4.1.
- The adjusted value of an observation is given by

  adjusted value = observed response value
                    + estimated regression coefficient
                    × (grand mean of covariate
                    − observed covariate value).

TABLE 4.23
Blood Pressure, Family History, and Smoker Status

| Family History | Smoking Status | | |
	Nonsmoker	Exsmoker	Current Smoker
Yes	125	114	135
	156	107	120
	103	134	123
	129	140	113
	110	120	165
	128	115	145
	135		120
No	114	110	140
	110	128	125
	91	105	123
	136	90	108
	105		113
	125		160
	103		
	110		

TABLE 4.24
Data from a Trial of Estrogen Patches in the Treatment of Postnatal Depression

Treatment	Baseline 1	Baseline 2	Depression Score
Placebo	18	18	15
	25	27	10
	24	17	12
	19	15	5
	22	20	5
	27	28	9
	21	16	11
	26	26	13
	20	19	6
	24	20	18
	24	22	10
Active	27	27	7
	19	15	8
	25	28	2
	19	18	6
	21	20	11
	21	21	5
	25	24	11
	25	25	6
	15	22	6
	27	26	10

mother's depression. Do these plots give you any concerns about the results of the analysis reported in Table 4.16? If so, reanalyze the data as you see fit.

4.6. Consider different orders of main effects in the multivariate analysis described in Section 4.7.

4.7. Plot some suitable graphs to aid the interpretation of the origin $\times$ sex $\times$ type interaction found in the analysis of the data on Australian school children.

4.8. Graphically explore the relationship between errors and the distraction measure in the smoking and performance data in Table 4.17. Do your graphs cause you any concern about the validity of the ANCOVA reported in Section 4.6?

4.9. Use the formula given in Display 4.4 to calculate the "adjusted" values of the observations for the pattern recognition experiment. Using these values, plot a suitable diagram to aid in the interpretation of the significant group $\times$ task interaction.

4.10. The data in Table 4.23 (taken from Boniface, 1995) were collected during a survey of systolic blood pressure of individuals classed according to smoking status and family history of circulation and heart problems. Carry out an analysis of variance of the data and state your conclusions. Examine the residuals from fitting what you consider the most suitable model and use them to assess the assumptions of your analysis of variance.

4.11. In the model for a balanced two-way fixed effects design (see Display 4.1), suggest sensible estimators for the main effect and interaction parameters.

4.12. The observations in Table 4.24 are part of the data collected in a clinical trial of the use of estrogen patches in the treatment of postnatal depression. Carry out an analysis of variance of the posttreatment measure of depression, using both pretreatment values as covariates.

5

Analysis of Repeated Measure Designs

5.1. INTRODUCTION

Many studies undertaken in the behavioral sciences and related disciplines involve recording the value of a response variable for each subject under more than one condition, on more than one occasion, or both. The subjects may often be arranged in different groups, either those naturally occurring such as gender, or those formed by random allocation, for example, when competing therapies are assessed. Such a design is generally referred to as involving *repeated measures*. Three examples will help in getting a clearer picture of a repeated measures type of study.

1. *Visual acuity data.* This example is one already introduced in Chapter 2, involving response times of subjects using their right and left eyes when light was flashed through lenses of different powers. The data are given in Table 2.3. Here there are two *within subject factors*, eye and lens strength.

2. *Field dependence and a reverse Stroop task.* Subjects selected randomly from a large group of potential subjects identified as having field-independent or field-dependent cognitive style were required to read two types of words (color and form names) under three cue conditions—normal, congruent, and incongruent.

TABLE 5.1

Field Independence and a Reverse Stroop Task

	b_1 Form			b_2 Color		
Subject	c_1 (N)	c_2 (C)	c_3 (I)	c_1 (N)	c_2 (C)	c_3 (I)
a_1 (Field Independent)						
1	191	206	219	176	182	196
2	175	183	186	148	156	161
3	166	165	161	138	146	150
4	206	190	212	174	178	184
5	179	187	171	182	185	210
6	183	175	171	182	185	210
7	174	168	187	167	160	178
8	185	186	185	153	159	169
9	182	189	201	173	177	183
10	191	192	208	168	169	187
11	162	163	168	135	141	145
12	162	162	170	142	147	151
a_2 (Field Dependent)						
13	277	267	322	205	231	255
14	235	216	271	161	183	187
15	150	150	165	140	140	156
16	400	404	379	214	223	216
17	183	165	187	140	146	163
18	162	215	184	144	156	165
19	163	179	172	170	189	192
20	163	159	159	143	150	148
21	237	233	238	207	225	228
22	205	177	217	205	208	230
23	178	190	211	144	155	177
24	164	186	187	139	151	163

Note. Response variable is time in milliseconds. N, normal; C, congruent; I, incongruent.

The dependent variable was the time in milliseconds taken to read the stimulus words. The data are shown in Table 5.1. Here there are two within subjects factors, type and cue, and one *between subjects factor*, cognitive style.

3. *Alcohol dependence and salsolinol excretion.* Two groups of subjects, one with severe dependence and one with moderate dependence on alcohol, had their salsolinol excretion levels (in millimoles) recorded on four consecutive days (for those readers without the necessary expertise in chemistry, salsolinol is an alkaloid

TABLE 5.2

Salsolinol Excretion Rates (mmol) for Moderately and Severely Dependent Alcoholic Patients

Subject	Day			
	1	2	3	4
Group 1 (Moderate Dependence)				
1	0.33	0.70	2.33	3.20
2	5.30	0.90	1.80	0.70
3	2.50	2.10	1.12	1.01
4	0.98	0.32	3.91	0.66
5	0.39	0.69	0.73	2.45
6	0.31	6.34	0.63	3.86
Group 2 (Severe Dependence)				
	1	2	3	4
7	0.64	0.70	1.00	1.40
8	0.73	1.85	3.60	2.60
9	0.70	4.20	7.30	5.40
10	0.40	1.60	1.40	7.10
11	2.50	1.30	0.70	0.70
12	7.80	1.20	2.60	1.80
13	1.90	1.30	4.40	2.80
14	0.50	0.40	1.10	8.10

with a structure similar to heroin). Primary interest centers on whether the groups behaved differently over time. The data are given in Table 5.2. Here there is a single within subject factor, time, and one between subjects factor, level of dependence.

Researchers typically adopt the repeated measures paradigm as a means of reducing error variability and/or as the natural way of measuring certain phenomena (e.g., developmental changes over time, and learning and memory tasks). In this type of design, effects of experimental factors giving rise to the repeated measures are assessed relative to the average response made by a subject on all conditions or occasions. In essence, each subject serves as his or her own control, and, accordingly, variability caused by differences in average responsiveness of the subjects is eliminated from the extraneous error variance. A consequence of this is that the power to detect the effects of within subjects experimental factors is increased compared to testing in a between subjects design.

Unfortunately, the advantages of a repeated measures design come at a cost, and that is the probable lack of independence of the repeated measurements. Observations made under different conditions involving the same subjects will very likely be correlated rather than independent. You will remember that in Chapters 3 and 4 the independence of the observations was one of the assumptions underlying the techniques described, so it should come as no great surprise that accounting for the dependence between observations in repeated measure designs brings about some problems in the analysis of such designs, as we shall see later.

The common feature of the three examples introduced in this section is that the subjects have the response variable recorded more than once. In the field-dependence example, the repeated measures arise from observing subjects under two different factors—the within subject factors. In addition, the subjects can be divided into two groups: the two levels of the between group factor. In the visual acuity example, only within subject factors occur. In both these examples it is possible (indeed likely) that the conditions under which a subject is observed are given in a random order. In the third example, however, where time is the single within subject factor, randomization is not, of course, an option. This makes the type of study where subjects are simply observed over time rather different from other repeated measures designs, and they are often given a different label, *longitudinal designs*. Because of their different nature, we shall consider them in a separate chapter (Chapter 7).

In a longitudinal design, the repeated measures on the same subject are a necessary part of the design. In other examples of repeated measures, however, it would be quite possible to use different groups of subjects for each condition combination, giving rise to a factorial design as discussed in the previous chapter. The primary purpose of using a repeated measures design in such cases is the control that the approach provides over individual differences between subjects. In the area of behavioral sciences, such differences are often quite large relative to differences produced by manipulation of the experimental conditions or treatments that the investigator is trying to evaluate.

5.2. ANALYSIS OF VARIANCE
FOR REPEATED MEASURE DESIGNS

The structure of the visual acuity study and the field-dependence study is, superficially at least, similar to that of the factorial designs of Chapter 4, and it might be imagined that the analysis of variance procedures described there could again be applied here. This would be to overlook the fundamental difference that in a repeated measures design the *same* subjects are observed under the levels of at least some of the factors of interest, rather than a *different* sample as required in a factorial design. It *is* possible, however, to use relatively straightforward ANOVA procedures for repeated measures data if three particular assumptions about the observations are valid. They are as follows.

1. *Normality*: the data arise from populations with normal distributions.
2. *Homogeneity of variance*: the variances of the assumed normal distributions are required to be equal.
3. *Sphericity*: the variances of the differences between all pairs of repeated measurements are equal. This condition implies that the correlations between pairs of repeated measures are also equal, the so-called *compound symmetry* pattern. Sphericity must also hold in all levels of the between subjects part of the design for the analyses to be described in Subsections 5.2.1 and 5.2.2 to be strictly valid.

We have encountered the first two requirements in previous chapters, and they need not be further discussed here. The sphericity condition has not, however, been met before, and it is particularly critical in the ANOVA of repeated measures data; the condition will be discussed in detail in Section 5.3. For the moment, however, let us assume that this is the best of all possible worlds, in which data are always normal, variances of different populations are always equal, and repeated measures always meet the sphericity requirement.

5.2.1. Repeated Measures Model for Visual Acuity Data

The results of a repeated measures ANOVA for the visual acuity data are shown in Table 5.3. The model on which this analysis is based is described in Display 5.1. The model allows the repeated measurements to be correlated but only to the extent of the compound symmetry pattern to be described in Section 5.3. It appears that only lens strength affects the response variable in this experiment. The means for

TABLE 5.3
ANOVA of Visual Acuity Data

Source	SS	DF	MS	F	p
Eye	52.90	1	52.90	0.61	.44
Error	1650.60	19	86.87		
Lens strength	571.85	3	190.62	12.99	<.001
Error	836.15	57	14.67		
Eye × Strength	43.75	3	14.58	1.51	.22
Error	551.75	57	9.68		

Display 5.1

Repeated Measures ANOVA Model for Two-Factor Design (A and B) with Repeated Measures on Both Factors

---

- Let y_{ijk} represent the observation on the ith subject for factor A level j and factor B level k. The model for the observations is

$$y_{ijk} = \mu + \alpha_j + \beta_k + \gamma_{jk} + \tau_i + (\tau\alpha)_{ij} + (\tau\beta)_{ik} + (\tau\gamma)_{ijk} + \epsilon_{ijk},$$

where α_j represents the effect of the jth level of factor A (with a levels), β_k is the effect of the kth level of factor B (with b levels), and γ_{jk} the interaction effect of the two factors. The term τ_i is a constant associated with subject i, and $(\tau\alpha)_{ij}$, $(\tau\beta)_{ik}$, and $(\tau\gamma)_{ijk}$ represent interaction effects of subject i with each factor and their interaction. As usual, ϵ_{ijk} represent random error terms. Finally, we assume that there are n subjects in the study (see Maxwell and Delaney, 1990, for full details).
- The factor A (α_j), factor B (β_k), and the AB interaction (γ_{jk}) terms are assumed to be fixed effects, but the subject and error terms are assumed to be random variables from normal distributions with zero means and variances specific to each term.
- This is an example of a *mixed model*. Correlations between the repeated measures arise from the subject terms because these apply to each of the measurements made on the subject.
- The analysis of variance table is as follows:

Source	SS	DF	MS	MSR(F)
A	ASS	$a-1$	$\frac{\text{ASS}}{(a-1)}$	AMS/EMS1
A × Subjects (error 1)	ESS1	$(a-1)(n-1)$	$\frac{\text{ESS1}}{(a-1)(n-1)}$	
B	BSS	$b-1$	$\frac{\text{BSS}}{(b-1)}$	BMS/EMS2
B × Subjects (error 2)	ESS2	$(b-1)(n-1)$	$\frac{\text{ESS2}}{(b-1)(n-1)}$	
AB	ABSS	$(a-1)(b-1)$	$\frac{\text{ABSS}}{(a-1)(b-1)}$	ABMS/EMS3
AB × Subjects (error 3)	ESS3	$(a-1)(b-1)(n-1)$	$\frac{\text{ESS3}}{(a-1)(b-1)(n-1)}$	

- Here it is useful to give specific expressions both for the various sums of squares above and for the terms that the corresponding mean squares are estimating (the *expected mean squares*), although the information looks a little daunting!

Source	Expression for SS	Expected Mean Square
A	$bn\sum_{j=1}^{n}(\bar{y}_{.j.} - \bar{y}_{...})^2$	$\sigma_\epsilon^2 + b\sigma_{\tau\alpha}^2 + nb\theta_\alpha^2$
A × S	$b\sum_{j=1}^{a}\sum_{i=1}^{n}(\bar{y}_{ij.} - \bar{y}_{i..} - y_{.j.} + \bar{y}_{...})^2$	$\sigma_\epsilon^2 + b\sigma_{\tau\alpha}^2$
B	$an\sum_{k=1}^{b}(\bar{y}_{..k} - \bar{y}_{...})^2$	$\sigma_\epsilon^2 + a\sigma_{\tau\beta}^2 + na\theta_\beta^2$
B × S	$a\sum_{k=1}^{b}\sum_{i=1}^{n}(\bar{y}_{i.k} - \bar{y}_{i..} - \bar{y}_{..k} - \bar{y}_{...})^2$	$\sigma_\epsilon^2 + a\sigma_{\tau\beta}^2$
A × B	$n\sum_{j=1}^{a}\sum_{k=1}^{b}(\bar{y}_{.jk} - \bar{y}_{.j.} - \bar{y}_{..k} + \bar{y}_{...})^2$	$\sigma_\epsilon^2 + \sigma_{\tau\gamma}^2 + n\theta_\gamma^2$
A × B × S	$\sum_{k=1}^{b}\sum_{j=1}^{a}\sum_{i=1}^{n}(y_{ijk} - \bar{y}_{ij.} - \bar{y}_{i.k}$ $-\bar{y}_{.jk} + \bar{y}_{i..} + \bar{y}_{.j.} + \bar{y}_{..k} - \bar{y}_{...})^2$	$\sigma_\epsilon^2 + \sigma_{\tau\gamma}^2$

(Continued)

Display 5.1

(Continued)

Here S represents subjects, the sigma terms are variances of the various random effects in the model, identified by an obvious correspondence to these terms, and $\theta_\alpha^2 = \sum_{j=1}^{a} \alpha_j^2$, $\theta_\beta^2 = \sum_{k=1}^{b} \beta_k^2$, $\theta_\gamma^2 = \sum_{j=1}^{a} \sum_{k=1}^{b} \gamma_{jk}^2$ are A, B, and AB effects that are nonzero unless the corresponding null hypothesis that the α_j, β_k, or γ_{jk} terms are zero is true.

• F tests of the various effects involve different error terms essentially for the same reason as discussed in Chapter 4, Section 4.5, in respect to a simple random effects model. For example, under the hypothesis of no factor A effect, the mean square for A and the mean square for A × Subjects are both estimators of the same variance. If, however, the factor effects α_j are not zero, then θ_α^2 is greater than zero and the mean square for A estimates a larger variance than the mean square for A × Subjects. Thus the ratio of these two mean squares provides an F test of the hypothesis that the populations corresponding to the levels of A have the same mean. (Maxwell and Delaney, 1990, give an extended discussion of this point.)

the four lens strengths are as follows.

6/6	6/18	6/36	6/60
112.93	115.05	115.10	118.23

It may be of interest to look in more detail at the differences in these means because the lens strengths form some type of ordered scale (see Exercise 5.2).

5.2.2. Model for Field Dependence Example

A suitable model for the field dependence example is described in Display 5.2. Again the model allows the repeated measurements to be related, but only to the extent of having the compound symmetry pattern. (The effects of departures from this pattern will be discussed later.) Application of the model to the field-dependence data gives the analysis of variance table shown in Table 5.4. The main effects of type and cue are highly significant, but a detailed interpretation of the results is left as an exercise for the reader (see Exercise 5.1).

5.3. SPHERICITY AND COMPOUND SYMMETRY

The analysis of variance of the two repeated measures examples described above are valid only if the normality, homogeneity, and sphericity assumptions are valid for the data. It is the latter that is of greatest importance in the repeated measures situation, because if the assumption is not satisfied, the F tests used are positively biased, leading to an increase in the probability of rejecting the null

Display 5.2

Repeated Measures ANOVA Model for Three-Factor Design (A, B, and C) with Repeated Measures on Two Factors (A and B)

- Let y_{ijkl} represent the observation on the ith person for factor A level j, factor B level k, and factor C level l. The model for the observations is

$$y_{ijkl} = \mu + \alpha_j + \beta_k + \gamma_{jk} + \tau_i + (\tau\alpha)_{ij} + (\tau\beta)_{ik} + (\tau\gamma)_{ijk}$$
$$+ \delta_l + \omega_{jl} + \theta_{kl} + \pi_{jkl} + \epsilon_{ijkl},$$

where the terms are as in the model in Display 5.1, with the addition of parameters for the between subjects factor C (with c levels) and its interactions with the within subject factors and with their interaction (δ_l, ω_{jl}, θ_{kl}, π_{jkl}).
- Note that there are no subject effects involving factor C, because subjects are not crossed with this factor—it is a between subjects factor.
- Again the factor A, factor B, and factor C terms are assumed to be fixed effects, but the subject terms and error terms are assumed to be random effects that are normally distributed with mean zero and particular variances.
- The analysis of variance table is as follows.

Source	SS	DF	MS	MSR(F)
Between subjects		$n-1$		
C	CSS	$c-1$	$\frac{\text{CSS}}{(c-1)}$	CMS/EMS1
Subjects within C (error 1)	ESS1	$n-c$	$\frac{\text{ESS1}}{(n-c)}$	
Within subjects		$n(a-1)$ $(b-1)$		
A	ASS	$a-1$	$\frac{\text{ASS}}{(a-1)}$	AMS/EMS2
C × A	CASS	$(c-1)(a-1)$	$\frac{\text{CASS}}{(c-1)(a-1)}$	CAMS/EMS2
C × A × Subjects within C (error 2)	ESS2	$(n-c)(a-1)$	$\frac{\text{ESS2}}{(n-c)(a-1)}$	
B	BSS	$b-1$	$\frac{\text{BSS}}{(b-1)}$	BMS/EMS3
C × B	CBSS	$(c-1)(b-1)$	$\frac{\text{CBSS}}{(c-1)(b-1)}$	CBMS/EMS3
C × B × Subjects within C (error 3)	ESS3	$(n-c)(b-1)$	$\frac{\text{ESS3}}{(n-c)(b-1)}$	
AB	ABSS	$(a-1)(b-1)$	$\frac{\text{ABSS}}{(a-1)(b-1)}$	ABMS/EMS4
C × A × B	CABSS	$(c-1)$ $(a-1)(b-1)$	$\frac{\text{CABSS}}{(c-1)(a-1)(b-1)}$	CABMS/EMS4
C × A × B × Subjects within C (error 4)	ESS4	$(n-c)$ $(a-1)(b-1)$	$\frac{\text{ESS4}}{(n-c)(a-1)(b-1)}$	

- n is the total number of subjects; that is, $n = n_1 + n_2 + \cdots + n_c$, where n_l is the number of subjects in level l of the between subjects factor C.

TABLE 5.4
ANOVA for Reverse Stroop Experiment

Source	SS	DF	MS	F	p
Groups (G)	18906.25	1	18906.25	2.56	.12
Error	162420.08	22	7382.73		
Type (T)	25760.25	1	25760.25	12.99	.0016
G × T	3061.78	1	3061.78	1.54	.23
Error	43622.30	22	1982.83		
Cue (C)	5697.04	2	2848.52	22.60	<.0001
G × C	292.62	2	146.31	1.16	.32
Error	5545.00	44	126.02		
T × C	345.37	2	172.69	2.66	.08
G × T × C	90.51	2	45.26	0.70	.50
Error	2860.79	44	65.02		

hypothesis when it is true; that is, increasing the size of the Type I error over the nominal value set by the experimenter. This will lead to an investigator's claiming a greater number of "significant" results than are actually justified by the data.

Sphericity implies that the variances of the differences between any pair of the repeated measurements are the same. In terms of the covariance matrix of the repeated measures, Σ (see the glossary in Appendix A), it requires the compound symmetry pattern, that is, the variances on the main diagonal must equal one another, and the covariances off the main diagonal must also all be the same. Specifically, the covariance matrix must have the following form:

$$\Sigma = \begin{pmatrix} \sigma^2 & \rho\sigma^2 & \cdots & \rho\sigma^2 \\ \rho\sigma^2 & \sigma^2 & \cdots & \rho\sigma^2 \\ \vdots & \vdots & \vdots & \vdots \\ \rho\sigma^2 & \rho\sigma^2 & \cdots & \sigma^2 \end{pmatrix}, \tag{5.1}$$

where σ^2 is the assumed common variance and ρ is the assumed common correlation coefficient of the repeated measures. This pattern must hold in each level of the between subject factors—quite an assumption!

As we shall see in Chapter 7, departures from the sphericity assumption are very likely in the case of longitudinal data, but perhaps less likely to be so much of a problem in experimental situations in which the levels of the within subject factors are given in a random order to each subject in the study. But where departures are suspected, what can be done?

5.4. CORRECTION FACTORS
IN THE ANALYSIS OF VARIANCE
OF REPEATED MEASURE DESIGNS

Box (1954) and Greenhouse and Geisser (1959) considered the effects of departures from the sphericity assumption in a repeated measures ANOVA. They demonstrated that the extent to which a set of repeated measures data deviates from the sphericity assumption can be summarized in terms of a parameter, ϵ, which is a function of the covariances and variances of the repeated measures. (In some ways it is a pity that ϵ has become the established way to represent this correction term, as there is some danger of confusion with all the epsilons occurring as error terms in models. Try not to become too confused!) Furthermore, an estimate of this parameter can be used to *decrease* the degrees of freedom of F tests for the within subjects effects, to account for departure from sphericity. In this way larger values will be needed to claim statistical significance than when the correction is not used, and so the increased risk of falsely rejecting the null hypothesis is removed. For those readers who enjoy wallowing in gory mathematical details, the formula for ϵ is given in Display 5.3. When sphericity holds, ϵ takes its maximum value as one and the F tests need not be amended. The minimum value of ϵ is $1/(p - 1)$, where p is the number of repeated measures. Some authors, for example, Greenhouse and Geisser (1959), have suggested using this lower bound in all cases so as to avoid the need to estimate ϵ from the data (see below). Such an approach is, however, very conservative. That is, it will too often fail to reject the null hypothesis when it is false. The procedure is not recommended, particularly now that most software packages will estimate ϵ routinely. In fact, there have been two suggestions for such estimation, both of which are usually provided by major statistical software.

1. Greenhouse and Geisser (1959) suggest simply substituting sample values for the population quantities in Display 5.2.

Display 5.3
Greenhouse and Geisser Correction Factor

- The correction factor ϵ is given by

$$\epsilon = \frac{p^2(\bar{\sigma}_{tt} - \bar{\sigma})^2}{(p-1)[\sum\sum \sigma_{ts}^2 - 2p\sum \bar{\sigma}_t^2 + p^2\bar{\sigma}^2]},$$

where p is the number of repeated measures on each subject, $(\sigma_{ts}, t = 1, \ldots, p,\ s = 1, \ldots, p$ represents the elements of the population covariance matrix of the repeated measures, and the remaining terms are as follows: $\bar{\sigma}$; the mean of all the elements of the covariance matrix; $\bar{\sigma}_{tt}$; the mean of the elements of the main diagonal; $\bar{\sigma}_t$; the mean of the elements in row t. (Covariance matrices are defined in the glossary in Appendix A.)

---

2. Huynh and Feldt (1976) suggest taking the estimate of ϵ to be $\min(1, a/b)$, where $a = ng(p-1)\hat{\epsilon} - 2$ and $b = (p-1)[g(n-1) - (p-1)\hat{\epsilon}]$, and where n is the number of subjects in each group, g is the number of groups, and $\hat{\epsilon}$ is the Greenhouse and Geisser estimate.

To illustrate the use of the correction factor approach to repeated measures data we shall again use the field-dependence data in Table 5.1, but now we shall assume that the six repeated measures arise from measuring the response variable, time, under six different conditions, that is, C1, C2, C3, C4, C5, C6, given in a random order to each subject. The scatterplot matrices of the observations in each group are shown in Figures 5.1 (field independent), and 5.2 (field dependent). Although there are not really sufficient observations to make convincing claims, there does appear to be some evidence of a lack of compound symmetry in the data. This is reinforced by look at the variances of differences between pairs of observations and the covariance matrices of the repeated measures in each level of cognitive style shown in Table 5.5. Clearly, there are large differences between the groups in terms of variances and covariances that may have implications for analysis, but that we shall ignore here (but see Exercise 5.1).

Details of the calculation of both the Greenhouse and Geisser and Huynh and Feldt estimates of ϵ are given in Table 5.6, and the results of the repeated measures ANOVA of the data with and without using the correction factor are given in Table 5.7. Here use of the correction factors does not alter which effects are statistically significant, although p values have changed considerably.

5.5. MULTIVARIATE ANALYSIS OF VARIANCE FOR REPEATED MEASURE DESIGNS

An alternative approach to the use of correction factors in the analysis of repeated measures data, when the sphericity assumption is judged to be inappropriate, is to use a multivariate analysis of variance. This technique has already been considered briefly in Chapters 3 and 4, in the analysis of studies in which a series of different response variables are observed on each subject. However, this method can also be applied in the repeated measures situation, in which a single response variable is observed under a number of different conditions and/or at a number of different times. The main advantage of using MANOVA for the analysis of repeated measures designs is that no assumptions now have to be made about the pattern of covariances between the repeated measures. In particular, these covariances need not satisfy the compound symmetry condition. A disadvantage of using MANOVA for repeated measures is often stated to be the technique's relatively low power when the assumption of compound symmetry is actually valid. However, Davidson (1972) compares the power of the univariate and multivariate analysis of variance

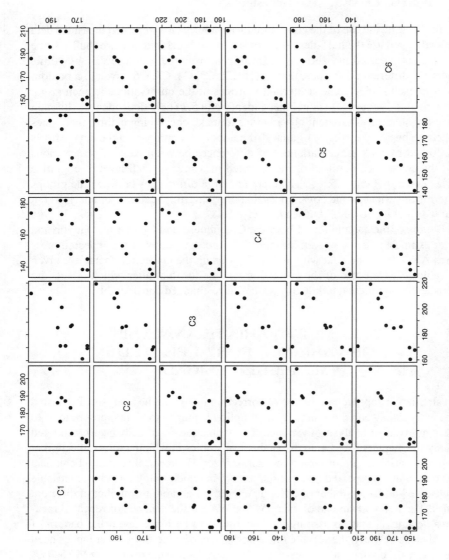

FIG. 5.1.   Scatterplot matrix of repeated measures for the field-independent group in the Stroop data.

144

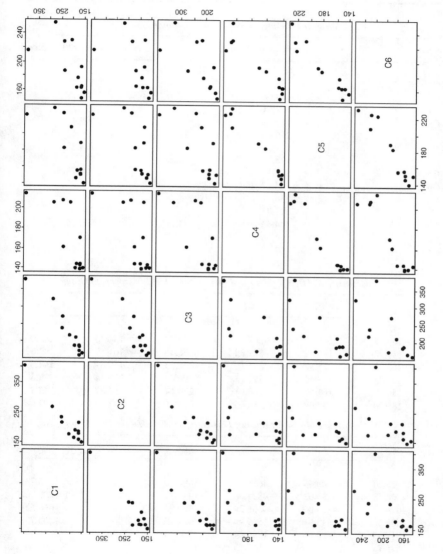

FIG. 5.2. Scatterplot matrix of repeated measures for the field-dependent group in the Stroop data.

145

TABLE 5.5
Covariance Matrices and Variances of Differences of Repeated Measures
for Stroop Data

Variances of Differences Between Conditions

Difference	Field-Independent Variance	Field-Dependent Variance
C1–C2	68.52	478.36
C1–C3	138.08	344.99
C1–C4	147.61	2673.36
C1–C5	116.02	2595.45
C1–C6	296.79	3107.48
C2–C3	123.36	778.27
C2–C4	178.00	2866.99
C2–C5	123.54	2768.73
C2–C6	304.27	3364.93
C3–C4	330.81	2461.15
C3–C5	311.42	2240.99
C3–C6	544.63	2500.79
C4–C5	16.08	61.54
C4–C6	47.00	150.97
C5–C6	66.27	103.66

Covariance Matrix for Repeated Measures for Field-Independent Group

	C1	C2	C3	C4	C5	C6
C1	169.15	145.73	203.12	160.82	154.70	185.00
C2	145.73	190.82	221.32	156.46	161.77	192.09
C3	203.12	221.32	375.17	172.23	160.01	164.09
C4	160.82	156.46	172.23	300.09	270.14	375.36
C5	154.70	161.77	160.01	270.14	256.27	343.82
C6	185.00	192.09	164.09	375.36	343.82	497.64

Covariance Matrix for Repeated Measures for Field-Dependent Group

	C1	C2	C3	C4	C5	C6
C1	5105.30	4708.48	4698.27	1698.36	1847.93	1595.73
C2	4708.48	4790.02	4324.00	1443.91	1603.66	1309.36
C3	4698.27	4324.00	4636.24	1569.94	1790.64	1664.55
C4	1698.36	1443.91	1569.94	964.79	1044.64	1003.73
C5	1847.93	1603.66	1790.64	1044.64	1186.02	1138.00
C6	1595.73	1309.36	1664.55	1003.73	1138.00	1193.64

TABLE 5.6
Calculating Correction Factors

The Mean of the Covariance Matrices in Table 5.5 is the Matrix, S Say, Given By

	C1	C2	C3	C4	C5	C6
C1	2637.22	2427.10	2450.70	929.59	1001.31	890.364
C2	2427.10	2490.42	2272.66	800.18	882.72	750.73
C3	2450.70	2272.66	2505.71	871.08	975.32	914.318
C4	929.59	800.18	871.08	632.44	657.39	689.55
C5	1001.31	882.72	975.32	657.39	721.14	740.91
C6	890.36	750.73	914.32	689.55	740.91	845.64

Mean of diagonal elements of **S** is 1638.76
Mean of all the elements in **S** is 1231.68
Sum of squares of all the elements in **S** is 72484646
The row means of **S** are

C1	C2	C3	C4	C5	C6
1722.72	1603.97	1664.96	763.37	829.80	805.25

Sum of the squared row means of **S** is 10232297

Greenhouse and Geisser Estimate (of Correction Factor)

$$\hat{\epsilon} = \frac{[36 \times (1638.76 - 1231.68)^2]}{5 \times (72484646 - 12 \times 10232297 + 36 \times 1231.86 \times 1231.86)} = 0.277.$$

Huynh–Feldt Estimate

$$a = 12 \times 5 \times 2 \times \hat{\epsilon} - 2,$$
$$b = 5 \times (2 \times 11 - 5 \times \hat{\epsilon}),$$
$$\min(1, a/b) = 0.303.$$

approaches when compound symmetry holds and concludes that the latter is nearly as powerful as the former when the number of observations exceeds the number of repeated measures by more than 20.

To illustrate the use of the MANOVA approach to the analysis of repeated measures data, we shall apply it to the field-dependence data, again assuming that the six repeated measures for each subject arise from measuring the reponse variable time under six conditions given in a random order. For the moment, forget the division of the data into two groups, field independent and field dependent, and assume that the null hypothesis of interest is whether the population means of the response variable differ for the six conditions. The multivariate approach to testing this hypothesis uses a version of Hotelling's T^2 statistic as described in Display 5.4. Numerical results for the Stroop data are given in Table 5.8. There is a highly significant effect of condition.

TABLE 5.7

ANOVA With and Without Correction Factors for Stroop Data

Source	SS	DF	MS	F	p
	Analysis of Variance Without Correction Factors				
Group (G)	17578.34	1	17578.34	2.38	.14
Error	162581.49	22	7390.07		
Condition (C)	28842.19	5	5768.42	11.81	<.00001
G × C	4736.28	5	947.26	1.94	.094
Error	53735.10	110	488.50		

Note. Repeated measures are assumed to arise from a single within subject factor with six levels. For these data the Greenhouse and Geisser estimate of the correction factor ϵ is 0.2768. The adjusted degrees of freedom and the revised p values for the within subject tests are as follows. Condition: df1 $= 0.2768 \times 5 = 1.38$; df2 $= 0.2768 \times 110 = 30.45$; $p = .0063$; Group × Condition: df1 $= 1.38$; df2 $= 30.45$; $p = .171$. The Huynh–Feldt estimate of ϵ is 0.3029 so that the following is true. Condition: df1 $= 0.3028 \times 5 = 1.51$; df2 $= 0.3029 \times 110 = 33.32$; $p = .00041$; Group × Condition: df1 $= 1.51$; df2 $= 33.32$; $p = .168$.

Display 5.4

Multivariate Test for Equality of Means of Levels of a Within Subject Factor A with a Levels

- The null hypothesis is that in the population, the means of the a levels of A are equal, that is,

$$H_0 : \mu_1 = \mu_2 = \cdots = \mu_a,$$

where $\mu_1, \mu_2, \ldots, \mu_a$ are the population means of the levels of A.
- This is equivalent to

$$H_0 : \mu_1 - \mu_2 = 0; \mu_2 - \mu_3 = 0; \cdots; \mu_{a-1} - \mu_a = 0$$

(see Exercise 5.5).
- Assessing whether these differences are *simultaneously* equal to zero gives the required multivariate test of H_0
- The appropriate test statistic is Hotelling's T^2 applied to the corresponding sample mean differences; that is; $\bar{\mathbf{d}}' = [\bar{x}_1 - \bar{x}_2, \bar{x}_2 - \bar{x}_3, \ldots, \bar{x}_{a-1} - \bar{x}_a]$.
- T^2 is given by

$$T^2 = n\bar{\mathbf{d}}'\mathbf{S_d}\bar{\mathbf{d}},$$

where n is the sample size and $\mathbf{S_d}$ is the covariance matrix of the differences between the repeated measurements.
- Under H_0, $F = (n - p + 1)/[(n - 1)(p - 1)]T^2$ has an F distribution with $p - 1$ and $n - p + 1$ degrees of freedom, where p is the number of repeated measures.

TABLE 5.8
Multivariate Test of Equality of Condition Means for Stroop Data

Subject	Differences Between Conditions				
	C1–C2	C2–C3	C3–C4	C4–C5	C5–C6
1	−15	−13	43	−6	−14
2	−8	−3	38	−8	−5
3	1	4	23	−8	−4
4	16	−22	38	−4	−6
5	−8	16	−11	−3	−25
6	8	4	−11	−3	−25
7	6	−19	20	7	−18
8	−1	1	32	−6	−10
9	−7	−12	28	−4	−6
10	−1	−16	40	−1	−18
11	−1	−5	33	−6	−4
12	0	−8	28	−5	−4
13	10	−55	117	−26	−24
14	19	−55	110	−22	−4
15	0	−15	25	0	−16
16	−4	25	165	−9	7
17	18	−22	47	−6	−17
18	−53	31	40	−12	−9
19	−16	7	2	−19	−3
20	4	0	16	−7	2
21	4	−5	31	−18	−3
22	28	−40	12	−3	−22
23	−12	−21	67	−11	−22
24	−22	−1	48	−12	−12

Note. The vector of means of these differences is $\bar{\mathbf{d}}' = [-1.42, -9.33, 40.88, -8.00, -10.92]$. The covariance matrix of the differences is

$$S_{\mathbf{d}} = \begin{pmatrix} 261.91 & -228.88 & 36.90 & 12.83 & -20.53 \\ -228.88 & 442.23 & -208.65 & 34.61 & 65.64 \\ 36.90 & -208.65 & 1595.51 & -143.13 & 100.10 \\ 12.83 & 34.61 & -143.13 & 54.52 & -14.70 \\ -20.53 & 65.64 & 100.10 & -14.70 & 81.73 \end{pmatrix}.$$

$T^2 = 115.08$, $F = 19.02$, and $p < .001$.

Display 5.5

Multivariate Test for an A × B Interaction in a Repeated Measures Design with Within Subject Factor A (at a Levels) and Between Subjects Factor B (at b Levels)

- Here the null can be expressed as follows:

$$H_0 : \mu_1^{(1)} - \mu_2(1) = \mu_1^{(2)} - \mu_2^{(2)}$$
$$\mu_2^{(1)} - \mu_3^{(1)} = \mu_2^{(2)} - \mu_3^{(2)}$$
$$\vdots$$
$$\mu_{a-1}^{(1)} - \mu_a^{(1)} = \mu_{a-1}^{(2)} - \mu_a^{(2)},$$

where $\mu_i^{(j)}$ represents the population mean of level i of A and level j of B. (The restriction to two levels for B is used simply of convenience of description.)

- Testing these equalities simultaneously gives the multivariate test for the A × B interaction.
- The relevant test is Hotelling's T^2 as described in Chapter 3, with the $\bar{\mathbf{x}}_1$ and $\bar{\mathbf{x}}_2$ terms being replaced by the vectors of group mean differences, $\mathbf{d}'_1 = [\bar{x}_1^{(1)} - \bar{x}_2^{(1)}, \bar{x}_2^{(1)} - \bar{x}_3^{(1)}, \ldots, \bar{x}_{a-1}^{(1)} - \bar{x}_a^{(1)}]$, and $\mathbf{d}_2$ defined similarly for the second group. Now $\mathbf{S}_1$ and $\mathbf{S}_2$ are the covariance matrices of the differences between the repeated measurements in each group.

Now what about the multivariate equivalent of the group × condition test? Again this involves Hotelling's T^2 test as introduced in Chapter 3 (see Display 3.8). Details are given in Display 5.5, and the numerical results for the Stroop data are given in Table 5.9. There is no evidence of a significant group × condition interaction.

Keselman, Keselman, and Lix (1995) give some detailed comparisons of univariate and multivariate approaches to the analysis of repeated measure designs, and, for balanced designs (between subjects factors), recommend the degrees of freedom adjusted univariate F test, as they suggest that the multivariate approach is affected more by departures from normality.

5.6. MEASURING RELIABILITY FOR QUANTITATIVE VARIABLES: THE INTRACLASS CORRELATION COEFFICIENT

One of the most important issues in designing experiments is the question of the reliability of the measurements. The concept of the reliability of a quantitative variable can best be introduced by use of statistical models that relate the unknown "true" values of a variable to the corresponding observed values. This is

TABLE 5.9

Multivariate Test for Groups × Condition Interaction for Stroop Data

---

• The covariance matrices of the condition differences for each group are

$$S_1 = \begin{pmatrix} 68.52 & -26.89 & -22.56 & 11.44 & 0.02 \\ -26.89 & 123.36 & -138.08 & -17.54 & -26.23 \\ -22.56 & -138.08 & 330.81 & -17.73 & 101.14 \\ 11.44 & -17.54 & -17.73 & 16.08 & -17.67 \\ 0.02 & -26.23 & 101.14 & -17.67 & 66.27 \end{pmatrix},$$

$$S_2 = \begin{pmatrix} 478.36 & -455.82 & 119.82 & 10.18 & -42.09 \\ -455.82 & 778.27 & -186.21 & 60.95 & 168.20 \\ 119.82 & -186.21 & 2461.15 & -140.85 & 85.18 \\ 10.18 & 60.95 & -140.85 & 61.54 & -7.11 \\ -42.09 & 168.21 & 85.18 & -7.11 & 103.66 \end{pmatrix}.$$

• The two vectors of the means of the differences are

$$\mathbf{d}_1' = [-0.83, -6.08, 25.08, -3.92, -11.58],$$

$$\mathbf{d}_2' = [-2.00, -12.58, 56.67, -12.08, -10.25].$$

• $T^2 = 11.83$, $F = 1.52$, $p = .23$.

---

a convenient point to introduce such models, because applying them generally involves a particular type of repeated measures design.

Examples of the models used for assessing reliability are given in Display 5.6; they all lead to the *intraclass correlation coefficient* as the way of indexing reliability. To illustrate the calculation of this coefficient, the ratings given by a number of judges to the competitors in a synchronized swimming competition will be used; these ratings are given in Table 5.10 (Of course, the author realizes that synchronized swimming may only be a minority interest among psychologists.) Before undertaking the necessary calculations, it might be useful to examine the data graphically in some way. Consequently, Figure 5.3 gives the box plots of the scores given by each judge, and Figure 5.4 shows the draughtsman's plot of the ratings made by each pair of judges. The box plots shown that the scores given by the first judge vary considerably more than those given by the other four judges. The scatterplots show that although there is, in general, a pattern of relatively strong relationships between the scores given by each pair of the five judges, this is not universally so; for example, the relationship between judges 4 and 5 is far less satisfactory.

The relevant ANOVA table for the synchronized swimming data is shown in Table 5.11, together with the details of calculating the intraclass correlation coefficient. The resulting value of 0.683 is not particularly impressive, and the

Display 5.6
Models for the Reliability of Quantitative Variables

- Let x represent the observed value of some variable of interest for a particular individual. If the observation was made a second time, say some days later, it would almost certainly differ to some degree from the first recording. A possible model for x is

$$x = t + \epsilon,$$

  where t is the underlying true value of the variable for the individual and ϵ is the measurement error.

- Assume that t has a distribution with mean μ and variance σ_t^2. In addition, assume that ϵ has a distribution with mean zero and variance σ_ϵ^2, and that t and ϵ are independent of each other; that is; the size of the measurement error does not depend on the size of the true value.

- A consequence of this model is that variability in the observed scores is a combination of true score variance and error variance.

- The reliability, R, of the measurements is defined as the ratio of the true score variance to the observed score variance,

$$R = \frac{\sigma_t^2}{\sigma_t^2 + \sigma_\epsilon^2},$$

  which can be rewritten as

$$R = \frac{1}{1 + \sigma_\epsilon^2/\sigma_t^2}.$$

- R is usually known as the *intraclass correlation coefficient*. As $\sigma_\epsilon^2/\sigma_t^2$ decreases, so that the error variance forms a decreasing part of the variability in the observations, R increases and its upper limit of unity is achieved when the error variance is zero.

- In the reverse case, where σ_ϵ^2 forms an increasing proportion of the observed variance, R decreases toward a lower limit of zero, which is reached when all the variability in the measurements results from the error component of the model.

- The intraclass correlation coefficient can be directly interpreted as the proportion of variance of an observation that is due to between subject variability in the true scores.

- When each of r observers rates a quantitative characteristic of interest on each of n subjects, the appropriate model becomes

$$x = t + o + \epsilon,$$

  where o represents the observer effect, which is assumed to be distributed with zero mean and variance σ_o^2.

- The three terms t, o, and ϵ are assumed to be independent of one another so that the variance of an observation is now

$$\sigma^2 = \sigma_t^2 + \sigma_o^2 + \sigma_\epsilon^2.$$

(Continued)

Display 5.6
(Continued)

- The interclass correlation coefficient for this situation is given by

$$R = \frac{\sigma_t^2}{\sigma_t^2 + \sigma_o^2 + \sigma_\epsilon^2}.$$

- An analysis of variance of the raters' scores for each subject leads to the following ANOVA table.

Source	DF	MS
Subjects	$n - 1$	SMS
Raters	$r - 1$	RMS
Error	$(n - 1)(r - 1)$	EMS

- It can be shown that the population or *expected* values of the three mean squares are

$$\text{SMS} : \sigma_\epsilon^2 + r\sigma_t^2,$$
$$\text{RMS} : \sigma_\epsilon^2 + n\sigma_o^2,$$
$$\text{EMS} : \sigma_\epsilon^2.$$

- By equating the observed values of the three mean squares to their expected values, the following estimators for the three variance terms σ_t^2, σ_o^2, and σ_ϵ^2 are found as follows:

$$\hat{\sigma}_t^2 = \frac{\text{SMS} - \text{EMS}}{r},$$
$$\hat{\sigma}_o^2 = \frac{\text{RMS} - \text{EMS}}{n},$$
$$\hat{\sigma}_\epsilon^2 = \text{EMS}.$$

- An estimator of R is then simply

$$\hat{R} = \frac{\hat{\sigma}_t^2}{\hat{\sigma}_t^2 + \hat{\sigma}_o^2 + \hat{\sigma}_\epsilon^2}.$$

synchronized swimmers involved in the competition might have some cause for concern over whether their performances were being judged fairly and consistently.

In the case of two raters giving scores on a variable to the same n subjects, the intraclass correlation coefficient is equivalent to Pearson's product moment correlation coefficient between $2n$ pairs of observations, of which the first n are the original values and the second n are the original values in reverse order. When only two raters are involved, the value of the intraclass correlation value depends in part on the corresponding product moment correlation coefficient and in part on the differences between the means and standard deviations of the two sets of

TABLE 5.10
Judges Scores for 40 Competitors in a Synchronized Swimming Competition

Swimmer	Judge 1	Judge 2	Judge 3	Judge 4	Judge 5
1	33.1	32.0	31.2	31.2	31.4
2	26.2	29.2	28.4	27.3	25.3
3	31.2	30.1	30.1	31.2	29.2
4	27.0	27.9	27.3	24.7	28.1
5	28.4	25.3	25.6	26.7	26.2
6	28.1	28.1	28.1	32.0	28.4
7	27.0	28.1	28.1	28.1	27.0
8	25.1	27.3	26.2	27.5	27.3
9	31.2	29.2	31.2	32.0	30.1
10	30.1	30.1	28.1	28.6	30.1
11	29.0	28.1	29.2	29.0	27.0
12	27.0	27.0	27.3	26.4	25.3
13	31.2	33.1	31.2	30.3	29.2
14	32.3	31.2	32.3	31.2	31.2
15	29.5	28.4	30.3	30.3	28.4
16	29.2	29.2	29.2	30.9	28.1
17	32.3	31.2	29.2	29.5	31.2
18	27.3	30.1	29.2	29.2	29.2
19	26.4	27.3	27.3	28.1	26.4
20	27.3	26.7	26.4	26.4	26.4
21	27.3	28.1	28.4	27.5	26.4
22	29.5	28.1	27.3	28.4	26.4
23	28.4	29.5	28.4	28.6	27.5
24	31.2	29.5	29.2	31.2	27.3
25	30.1	31.2	28.1	31.2	29.2
26	31.2	31.2	31.2	31.2	30.3
27	26.2	28.1	26.2	25.9	26.2
28	27.3	27.3	27.0	28.1	28.1
29	29.2	26.4	27.3	27.3	27.3
30	29.5	27.3	29.2	28.4	28.1
31	28.1	27.3	29.2	28.1	29.2
32	31.2	31.2	31.2	31.2	28.4
33	28.1	27.3	27.3	28.4	28.4
34	24.0	28.1	26.4	25.1	25.3
35	27.0	29.0	27.3	26.4	28.1
36	27.5	27.5	24.5	25.6	25.3
37	27.3	29.5	26.2	27.5	28.1
38	31.2	30.1	27.3	30.1	29.2
39	27.0	27.5	27.3	27.0	27.3
40	31.2	29.5	30.1	28.4	28.4

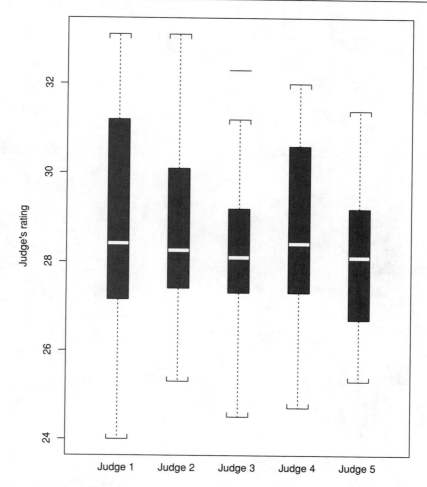

FIG. 5.3. Boxplots of scores given to synchronized swimming competitors by five judges.

ratings. The relationship between the intraclass correlation coefficient and Pearson's coefficient is given explicitly in Display 5.7.

Sample sizes required in reliability studies concerned with estimating the intraclass correlation coefficient are discussed by Donner and Eliasziw (1987), and the same authors (Eliasziw and Donner, 1987) also consider the question of the optimal choice of r and n that minimizes the overall cost of a reliability study. Their conclusion is that an increase in r for fixed n provides more information than an increase in n for fixed r.

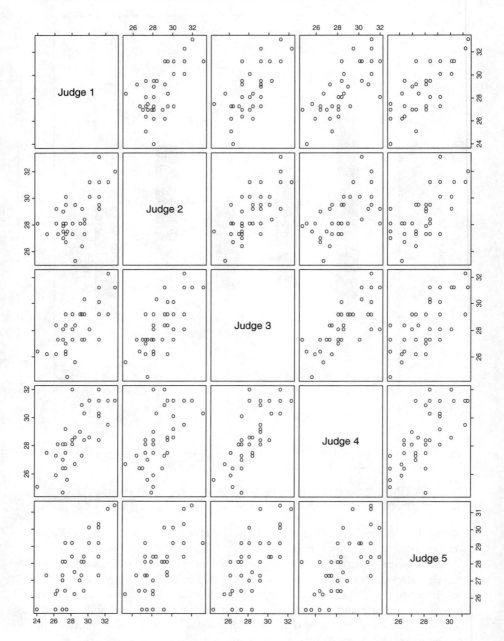

FIG. 5.4. Matrix of scatterplots for scores given by each pair of judges in rating synchronized swimmers.

TABLE 5.11
ANOVA and Calculation of Intraclass Correlation Coefficient
for Synchronized Swimming Judgments

| Source | ANOVA Table | | |
	SS	DF	MS
Swimmers	521.27	39	13.37
Judges	19.08	4	4.77
Error	163.96	156	1.05

Note. The required estimates of the three variance terms in the model are

$$\hat{\sigma}_r^2 = \frac{13.37 - 1.05}{5} = 2.53,$$
$$\hat{\sigma}_o^2 = \frac{4.77 - 1.05}{40} = 0.093,$$
$$\hat{\sigma}_\epsilon^2 = 1.05.$$

Consequently, the estimate of the intraclass correlation coefficient is

$$\hat{R} = \frac{2.53}{2.53 + 0.093 + 1.05} = 0.69.$$

Display 5.7
Relationship Between the Intraclass Correlation Coefficient, R, and the Product Moment
Correlation Coefficient, R_M, in the Case of Two Raters

$$R = \frac{[s_1^2 + s_2^2 - (s_1 - s_2)^2]R_M - (\bar{x}_1 - \bar{x}_2)^2/2}{s_1^2 + s_2^2 + (\bar{x}_1 - \bar{x}_2)^2/2},$$

where $\bar{x}_1$ and $\bar{x}_2$ are the mean values of the scores of the two raters, and s_1^2 and s_2^2 are
the corresponding variances.

5.7. SUMMARY

1. Repeated measures data arise frequently in many areas of psychological research and require special care in their analysis.
2. Univariate analysis of variance of repeated measures designs depends on the assumption of sphericity. This assumption is unlikely to be valid in most situations.
3. When sphericity does not hold, either a correction factor approach or a MANOVA approach can be used (other possibilities are described by Goldstein, 1995).

4. Recently developed methods for the analysis of repeated measure designs having a nonnormal response variable are described in Davis (1991).

COMPUTER HINTS

SPSS

In SPSS the basic steps to conduct a repeated measures analysis are as follows.

1. Click **Statistics**, click **General Linear Model**, and then click **GLM-Repeated Measures** to get the GLM-Repeated Measures Define Variable(s) dialog box.
2. Specify the names of the within subject factors, and the number of levels of these factors. Click **Add**.
3. Click **Define** and move the names of the repeated measure variables to the Within Subjects Variables box.

Both univariate tests, with and without correction factors, and multivariate tests can be obtained. A test of sphericity is also given.

TABLE 5.12
Skin Resistance Data

Subject	Electrode Type				
	1	2	3	4	5
1	500	400	98	200	250
2	660	600	600	75	310
3	250	370	220	250	220
4	72	140	240	33	54
5	135	300	450	430	70
6	27	84	135	190	180
7	100	50	82	73	78
8	105	180	32	58	32
9	90	180	220	34	64
10	200	290	320	280	135
11	15	45	75	88	80
12	160	200	300	300	220
13	250	400	50	50	92
14	170	310	230	20	150
15	66	1000	1050	280	220
16	107	48	26	45	51

TABLE 5.13
Data from Quitting Smoking Experiment

Gender	Treatment	Before		After	
		Home	Work	Home	Work
Male	Taper	7	6	6	4
		5	4	5	2
		8	7	7	4
		8	8	6	5
		6	5	5	3
	Intermediate	8	7	7	6
		5	5	5	4
		7	6	6	5
		8	7	6	5
		7	6	5	4
		5	6	6	6
		8	5	5	8
	Aversion	9	8	5	4
		4	4	3	2
		7	7	5	3
		7	5	5	0
		8	7	6	3
		9	5	4	2
Female	Taper	6	6	6	3
		4	5	6	5
		8	8	8	5
	Intermediate	7	7	6	5
		5	5	8	4
		6	6	6	6
		5	5	8	5
		6	9	6	8
		3	3	5	4
		5	5	6	8
	Aversion	7	8	8	5
		4	4	4	3
		9	9	6	3
		5	5	4	4
		6	8	6	8
		4	5	5	5
		3	3	4	6
		5	5	6	3
		5	6	8	7
		3	4	4	5

S-PLUS

In S-PLUS, the main function for analyzing repeated measures is *lme*, but because this is perhaps more applicable to longitudinal studies than to the examples covered in this chapter, we shall not say more about it until Chapter 7.

EXERCISES

5.1. Apply both the correction factor approach and the MANOVA approach to the Stroop data, now keeping both within subject factors. How do the results compare with the simple univariate ANOVA results given in the text? Do you think any observations should be removed prior to the analysis of these data?

5.2. Reanalyze the visual acuity data by using orthogonal polynomial contrasts for lens strength.

5.3. Five different types of electrodes were applied to the arms of 16 subjects and the resistance was measured in kilohms. The results are given in Table 5.12. The experiment was carried out to see whether all electrode types performed similarly. Calculate the intraclass correlation coefficient of the electrodes. Are there any observations that you think may be having an undue influence on the results?

5.4. The data in Table 5.13 are taken from an investigation of cigarette smoking. Three different procedures for quitting smoking (tapering off, immediate stopping, and aversion therapy) were compared. Subjects were randomly allocated to each treatment and were asked to rate (on a 1- to 10-point scale) their desire to smoke "right now" in two different environments (home versus work) both before and after quitting. Carry out both univariate and multivariate analyses of the data, noting that the groups formed by the two between subjects factors, gender and treatment group, have different numbers of subjects.

5.5. Show the equivalence of the two forms of the null hypothesis given in Display 5.5.

6

Simple Linear Regression and Multiple Regression Analysis

6.1. INTRODUCTION

In Table 6.1 a small set of data appears giving the average vocabulary size of children at various ages. Is it possible (or indeed sensible) to try to use these data to construct a model for predicting the vocabulary size of children older than 6, and how should we go about it? Such questions serve to introduce one of the most widely used of statistical techniques, *regression analysis*. (It has to be admitted that the method is also often misused.) In very general terms, regression analysis involves the development and use of statistical techniques designed to reflect the way in which variation in an observed random variable changes with changing circumstances. More specifically, the aim of a regression analysis is to derive an equation relating a dependent and an explanatory variable, or, more commonly, several explanatory variables. The derived equation may sometimes be used solely for prediction, but more often its primary purpose is as a way of establishing the relative importance of the explanatory variable(s) in determining the response variable, that is, in establishing a useful model to describe the data. (Incidentally, the term *regression* was first introduced by Galton in the 19th Century to characterize a tendency toward mediocrity, that is, more average, observed in the offspring of parents.)

TABLE 6.1
The Average Oral Vocabulary Size of Children
at Various Ages

Age (Years)	Number of Words
1.0	3
1.5	22
2.0	272
2.5	446
3.0	896
3.5	1222
4.0	1540
4.5	1870
5.0	2072
6.0	2562

In this chapter we shall concern ourselves with regression models for a response variable that is continuous; in Chapter 10 we shall consider suitable regression models for categorical response variables.

No doubt most readers will have covered *simple linear regression* for a response variable and a single explanatory variable in their introductory statistics course. Nevertheless, at the risk of enduring a little boredom, it may be worthwhile reading the next section both as an *aide memoir* and as an initial step in dealing with the more complex procedures needed when several explanatory variables are considered, a situation to be discussed in Section 6.3.

6.2.   SIMPLE LINEAR REGRESSION

The essential components of the simple linear regression model involving a single explanatory variable are shown in Display 6.1. (Those readers who require a more detailed acoount should consult Daly, Hand, Jones, Lunn, and McConway, 1995.) Fitting the model to the vocabulary data gives the results shown in Table 6.2. A plot of the fitted line, its 95% confidence interval, and the original data are given in Figure 6.1. The confidence interval for the regression coefficient of age (513.35, 610.51), indicates that there is a very strong relationship between vocabulary size and age. Because the estimated regression coefficient is positive, the relationship is such that as age increases so does vocabulary size—all rather obvious even from a simple plot of the data! The estimated regression coefficient implies that the increase in average vocabulary size corresponding to an increase in age of a year is approximately 562 words.

Display 6.1

Simple Linear Regression Model

- The basic idea of simple linear regression is that the mean values of a response variable y lie on a straight line when plotted against values of an explanatory variable, x.
- The variance of y for a given value of x is assumed to be independent of x.
- More specifically, the model can be written as

$$y_i = \alpha + \beta x_i + \epsilon_i,$$

where $y_1, y_2, \ldots, y_n$ are the n observed values of the response variable, and $x_1, x_2, \ldots, x_n$ are the corresponding values of the explanatory variable.
- The $\epsilon_i, i = 1, \ldots, n$ are known as residual or error terms and measure how much an observed value, y_i, differs from the value predicted by the model, namely $\alpha + \beta x_i$.
- The two parameters of the model, α and β, are the *intercept* and *slope* of the line.
- Estimators of α and β are found by minimizing the sum of the squared deviations of observed and predicted values, a procedure known as *least squares*. The resulting estimators are

$$\hat{\alpha} = \bar{y} - \hat{\beta}\bar{x},$$

$$\hat{\beta} = \frac{\sum_{i=1}^{n}(x_i - \bar{x})(y_i - \bar{y})}{\sum_{i=1}^{n}(x_i - \bar{x})^2},$$

where $\bar{y}$ and $\bar{x}$ are the sample means of the response and explanatory variable, respectively.
- The error terms, ϵ_i, are assumed to be normally distributed with mean zero and variance σ^2.
- An estimator of σ^2 is s^2 given by

$$s^2 = \frac{\sum_{i=1}^{n}(y_i - \hat{y}_i)^2}{n - 2},$$

where $\hat{y}_i = \hat{\alpha} + \hat{\beta}x_i$ is the predicted value or *fitted value* of the response variable for an individual with explanatory variable value x_i.
- Estimators of the standard errors of the estimated slope and estimated intercepts are given by

$$s_{\hat{\alpha}} = \sqrt{\frac{s^2}{n} + \frac{\bar{x}^2 s^2}{\sum_{i=1}^{n}(x_i - \bar{x})^2}},$$

$$s_{\hat{\beta}} = \frac{s}{\sqrt{\sum_{i=1}^{n}(x_i - \bar{x})^2}}.$$

- Confidence intervals for, and tests of hypotheses about, the slope and intercept parameters can be constructed in the usual way from these standard error estimators.
- The variability of the response variable can be partitioned into a part that is due to regression on the explanatory variable $[\sum_{i=1}^{n}(\hat{y}_i - \bar{y})^2]$ and a residual

---

(Continued)

Display 6.1

(Continued)

$[\sum_{i=1}^{n}(y_i - \hat{y}_i)^2]$. The relevant terms are usually arranged in an analysis of variance table as follows.

Source	SS	DF	MS	F
Regression	RGSS	1	RGSS/1	MSRG/MSR
Residual	RSS	n-2	RSS/$(n-2)$	

- The F statistic provides a test of the hypothesis that the slope parameter, β, is zero.
- The residual mean square gives the estimate of σ^2.

TABLE 6.2

Results of Fitting a Simple Linear Regression Model to Vocabulary Data

Parameter Estimates

Coefficient	Estimate	SE
Intercept	−763.86	88.25
Slope	561.93	24.29

ANOVA Table

Source	SS	DF	MS	F
Regression	7294087.0	1	7294087.0	539.19
Residual	109032.2	8	1369.0	

It is possible to use the derived linear regression equation relating average vocabulary size to age, to predict vocabulary size at different ages. For example, for age 5.5 the prediction would be

$$\text{average vocabulary size} = -763.86 + 561.93 \times 5.5 = 2326.7. \quad (6.1)$$

As ever, an estimate of this kind is of little use without some measure of its variability; that is, a confidence interval for the prediction is needed. Some relevant formulas are given in Display 6.2, and in Table 6.3, predicted vocabulary scores for a number of ages and their confidence intervals (CIs) are given. Note that the confidence intervals become wider. That is, the prediction becomes less certain as the age at which a prediction is made departs further from the mean of the observed ages.

Thus the derived regression equation does allow predictions to be made, but it is of considerable importance to reflect a little on whether such predictions are really sensible in this situation. A little thought shows that they are not. Using the fitted equation to predict future vocabulary scores is based on the assumption that the observed upward trend in vocabulary scores with age will continue unabated

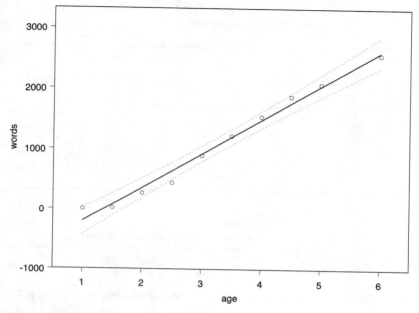

FIG. 6.1. Plot of vocabulary data, showing fitted regression and 95% confidence interval (dotted line).

Display 6.2
Using the Simple Linear Regression Model for Prediction

- The predicted response value corresponding to a value of the explanatory variable of say x_0 is

$$y_{\text{predicted}} = \hat{\alpha} + \hat{\beta}x_0.$$

- An estimator of the variance of a predicted value is provided by

$$s^2_{\text{predicted}} = s^2 \left[\frac{(x_0 - \bar{x})^2}{\sum_{i=1}^{n}(x_i - \bar{x})^2} + \frac{1}{n} + 1 \right].$$

(Note that the variance of the prediction increases as x_0 gets further away from $\bar{x}$.)
- A confidence interval for a prediction can now be constructed in the usual way:

$$y_{\text{predicted}} \pm s_{\text{predicted}} \times t_{n-2}(\alpha/2).$$

Here t is the value of t with $n-2$ degrees of freedom for the required $100(1-\alpha)\%$ confidence interval.

TABLE 6.3

Predictions and Their Confidence Intervals for the Vocabulary Data

Age	Predicted Vocabulary Size	SE of Prediction	95% CI
5.5	2326.7	133.58	(2018,2635)
7.0	3169.6	151.88	(2819,3520)
10.0	4855.4	203.66	(4385,5326)
20.0	10474.7	423.72	(9496,11453)

Display 6.3

Simple Linear Regression Model with Zero Intercept

- The model is now

$$y_i = \beta x_i + \epsilon_i.$$

- Application of least squares to this model gives the following estimator for β.

$$\hat{\beta} = \sum_{i=1}^{n} x_i y_i \Big/ \sum_{i=1}^{n} x_i^2.$$

- An estimator of the variance of $\hat{\beta}$ is given by

$$s_{\hat{\beta}}^2 = s^2 \Big/ \sum_{i=1}^{n} x_i^2,$$

where s^2 is the residual mean square from the relevant analysis of variance table.

into future ages at the same rate as observed between ages 1 and 6, namely between approximately 513 and 610 words per year. This assumption is clearly false, because the rate of vocabulary acquisition will gradually decrease. Extrapolation outside the range of the observed values of the explanatory variable is known in general to be a risky business, and this particular example is no exception.

Now you might remember that in Chapter 1 it was remarked that if you do not believe in a model you should not perform operations and analyses that assume it is true. Bearing this warning in mind, is the simple linear model fitted to the vocabulary data really believable? A little thought shows that it is not. The estimated vacabulary size at age zero, that is, the estimated intercept on the y, axis is -763.86, with an approximate 95% confidence interval of $(-940, -587)$. This is clearly a silly interval for a value that is known *a priori* to be zero. It reflects the inappropriate nature of the simple linear regression model for these data. An apparently more suitable model would be one in which the intercept is constrained to be zero. Estimation for such a model is described in Display 6.3. For the vocabulary data, the

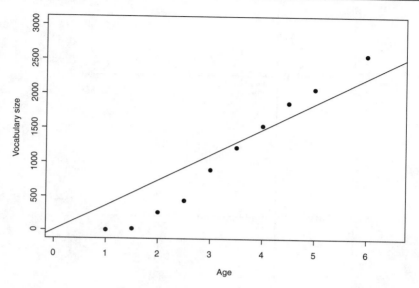

FIG. 6.2.   Plot of vocabulary data, showing the fitted zero intercept line.

estimated value of the regression coefficient is 370.96, with an estimated standard error of 30.84. The fitted zero intercept line and the original data are plotted in Figure 6.2. It is apparent from this plot that our supposedly more appropriate model does not fit the data as well as the one rejected as being unsuitable on logical grounds. So what has gone wrong? The answer is that the relationship between age and vocabulary size is more complex than is allowed for by either of the two regression models considered above.

One way of investigating the possible failings of a model was introduced in Chapter 4, namely the examination of residuals, that is, the difference between an observed value, y_i, and the value predicted by the fitted model, $\hat{y}_i$,

$$r_i = y_i - \hat{y}_i. \tag{6.2}$$

In a regression analysis there are various ways of plotting these values that can be helpful in assessing particular components of the regression model. The most useful plots are as follows.

1. A histogram or stem-and-leaf plot of the residuals can be useful in checking for symmetry and specifically for normality of the error terms in the regression model.
2. Plotting the residuals against the corresponding values of the explanatory variable. Any sign of curvature in the plot might suggest that, say, a quadratic term in the explanantory variable should be included in the model.

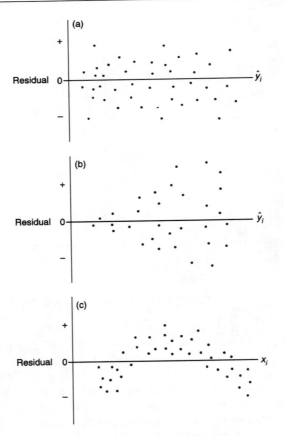

FIG. 6.3. Idealized residual plots.

3. Plotting the residuals against the *fitted values* of the response variable (not the response values themselves, for reasons spelled out in Rawlings, 1988). If the variability of the residuals appears to increase with the size of the fitted values, a transformation of the response variable prior to fitting is indicated.

Figure 6.3 shows some idealized residual plots that indicate particular points about models.

1. Figure 6.3(a) is what is looked for to confirm that the fitted model is appropriate.
2. Figure 6.3(b) suggests that the assumption of constant variance is not justified so that some transformation of the response variable before fitting might be sensible.

TABLE 6.4

Residuals from Fitting a Simple Linear Regression to the Vocabulary Data

Age	Vocabulary Size	Predicted Value	Residual
1.0	3	−201.93	204.93
1.5	22	79.03	−57.03
2.0	272	360.00	−87.99
2.5	446	640.96	−194.96
3.0	896	921.93	−25.92
3.5	1222	120.89	−19.11
4.0	1540	1483.86	56.15
4.5	1870	1764.82	105.19
5.0	2072	2045.79	26.22
6.0	2562	2607.72	−45.70

3. Figure 6.3(c) implies that the model requires a quadratic term in the explanantory variable.

(In practice, of course, the residual plots obtained might be somewhat more difficult to interpret than these idealized plots.)

Table 6.4 shows the numerical values of the residuals for the vocabulary data and Figure 6.4 shows the residuals plotted against age. Here there are very few observations on which to make convincing claims, but the pattern of residuals does seem to suggest that a more suitable model might be found for these data—see Exercise 6.1.

(The raw residuals defined and used above suffer from certain problems that make them less helpful in investigating fitted models than they might be. The problem will be taken up in detail in Section 6.5, where we also discuss a number of other regression diagnostics.)

6.3. MULTIPLE LINEAR REGRESSION

Multiple linear regression represents a generalization to more than a single explanatory variable, of the simple linear regression procedure described in the previous section. It is now the relationship between a response variable and several explanatory variables that becomes of interest. Details of the model, including the estimation of its parameters by least squares and the calculation of standard errors, are given in Display 6.4. (Readers to whom a matrix is still a mystery should avoid this display at all costs.) The regression coefficients in the

170

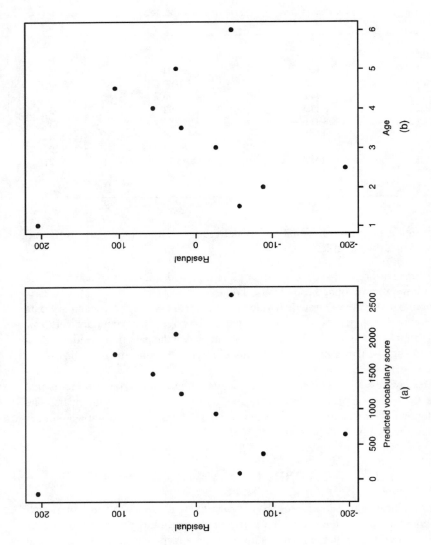

FIG. 6.4. Residuals from regression of vocabulary score on age, plotted (a) against predicted value and (b) against age.

Display 6.4
Multiple Regression Model

- The multiple linear regression model for a response variable y with observed values $y_1, y_2, \ldots, y_n$ and p explanatory variables, $x_1, x_2, \ldots, x_p$, with observed values $x_{i1}, x_{i2}, \ldots, x_{ip}$ for $i = 1, 2, \ldots, n$, is

$$y_i = \beta_0 + \beta_1 x_{i1} + \beta_2 x_{i2} + \cdots + \beta_p x_{ip} + \epsilon_i.$$

- The regression coefficients $\beta_1, \beta_2, \ldots, \beta_p$ give the amount of change in the response variable associated with a unit change in the corresponding explanatory variable, *conditional* on the other explanatory variables in the model remaining unchanged.
- The explanatory variables are strictly assumed to be fixed; that is, they are not random variables. In practice, where this is rarely the case, the results from a multiple regression analysis are interpreted as being *conditional* on the observed values of the explanatory variables.
- The "linear" in multiple linear regression refers to the parameters rather than the explanatory variables, so the model remains linear if, for example, a quadratic term, for one of these variables is included. (An example of a *nonlinear model* is $y = \beta_1 e^{\beta_2 x_1} + \beta_3 e^{\beta_4 x_2}$.)
- The residual terms in the model, $\epsilon_i, i = 1, \ldots, n$, are assumed to have a normal distribution with mean zero and variance σ^2. This implies that, for given values of the explanatory variables, the response variable is normally distributed with a mean that is a linear function of the explanatory variables and a variance that is not dependent on these variables.
- The aim of multiple regression is to arrive at a set of values for the regression coefficients that makes the values of the response variable predicted from the model as close as possible to the observed values.
- As in the simple linear regression model, the least-squares procedure is used to estimate the parameters in the multiple regression model.
- The resulting estimators are most conveniently written with the help of some matrices and vectors. The result might look complicated particularly if you are not very familiar with matrix algebra, but you can take my word that the result looks even more horrendous when written without the use of matrices and vectors!
- So by introducing a vector $\mathbf{y}' = [y_1, y_2, \ldots, y_n]$ and an $n \times (p + 1)$ matrix $\mathbf{X}$ given by

$$\mathbf{X} = \begin{pmatrix} 1 & x_{11} & x_{12} & \cdots & x_{1p} \\ 1 & x_{21} & x_{22} & \cdots & x_{2p} \\ \vdots & \vdots & \vdots & \vdots & \vdots \\ 1 & x_{n1} & x_{n2} & \cdots & x_{np} \end{pmatrix},$$

we can write the multiple regression model for the n observations concisely as

$$\mathbf{y} = \mathbf{X}\beta + \epsilon,$$

where $\epsilon' = [\epsilon_1, \epsilon_2, \ldots, \epsilon_n]$ and $\beta' = [\beta_0, \beta_1, \cdots, \beta_p]$.

(Continued)

Display 6.4

(Continued)

- The least-squares estimators of the parameters in the multiple regression model are given by the set of equations

$$\hat{\beta} = (\mathbf{X'X})^{-1}\mathbf{X'y}.$$

- These matrix manipulations are easily performed on a computer, but you must ensure that there are no linear relationships between the explanatory variables, for example, that one variable is the sum of several others; otherwise, your regression software will complain. (See Section 6.6.)

  (More details of the model in matrix form and the least-squares estimation process are given in Rawlings, 1988.)
- The variation in the response variable can be partitioned into a part due to regression on the explanatory variables and a residual as for simple linear regression. These can be arranged in an analysis of variance table as follows.

Source	DF	SS	MS	F
Regresssion	p	RGSS	RGSS/p	RGMS/RSMS
Residual	$n - p - 1$	RSS	RSS/$(n - p - 1)$	

- The residual mean square s^2 is an estimator of σ^2.
- The covariance matrix of the parameter estimates in the multiple regression model is estimated from

$$\mathbf{S}_{\hat{\beta}} = s^2(\mathbf{X'X})^{-1}.$$

  The diagonal elements of this matrix give the variances of the estimated regression coefficients and the off-diagonal elements their covariances.
- A measure of the fit of the model is provided by the *multiple correlation coefficient*, R, defined as the correlation between the observed values of the response variable, $y_1, \ldots, y_n$, and the values predicted by the fitted model, that is,

$$\hat{y}_i = \hat{\beta}_0 + \hat{\beta}_1 x_{i1} + \cdots + \hat{\beta}_p x_{ip}$$

- The value of R^2 gives the proportion of variability in the response variable accounted for by the explanatory variables.

---

multiple regression model give the change to be expected in the response variable when the corresponding explanatory variable changes by one unit, *conditional* on the other explanatory variables remaining unchanged. This is often referred to as 'partialling out' or 'controlling' for other variables, although such terms are probably best avoided. As in the simple linear regression model described in the previous section, the variability in the response variable in multiple regression can be partitioned into a part due to regression on the explanatory variables and a residual term. The resulting F-test can be used to assess the omnibus, and in most practical situations, relatively uninteresting null hypothesis, that *all* the regression coefficients are zero, i.e., none of the chosen explanatory variables are predictive of the response variable.

6.3.1. A Simple Example of Multiple Regression

As a gentle introduction to the topic, the multiple linear regression model will be applied to the data on ice cream sales introduced in Chapter 2 (see Table 2.4). Here the response variable is the consumption of ice cream measured over thirty 4-week periods. The explanatory variables believed to influence consumption are average price and mean temperature in the 4-week periods.

A scatterplot matrix of the three variables (see Figure 6.5) suggests that temperature is of most importance in determining consumption. The results of applying the multiple regression model to these data are shown in Table 6.5. The F value of 23.27 with 2 and 27 degrees of freedom has an associated p value that is very small. Clearly, the hypothesis that both regression coefficients are zero is not tenable.

For the ice cream data, the multiple correlation coefficient (defined in Display 6.4) takes the value 0.80, implying that the two explanantory variables, price and temperature, account for 64% of the variability in consumption. The negative regression coefficient for price indicates that, for a given temperature, consumption decreases with increasing price. The positive coefficient for temperature implies that, for a given price, consumption increases with increasing temperature. The sizes of the two regression coefficients might appear to imply that price is of more importance than temperature in predicting consumption. But this is an illusion produced by the different scales of the two variables. The raw regression coefficients should not be used to judge the relative importance of the explanatory variables, although the *standardized* values of these coefficients can, partially at least, be used in this way. The standardized values might be obtained by applying the regression model to the values of the response variable and explanatory variables, standardized (divided by) their respective standard deviations. In such an analysis, each regression coefficient represents the change in the standardized response variable, associated with a change of one standard deviation unit in the explanatory variable, again conditional on the other explanatory variables' remaining constant. The standardized regression coefficients can, however, be found without undertaking this further analysis, simply by multiplying the raw regression coefficient by the standard deviation of the appropriate explanatory variable and dividing by the standard deviation of the response variable. In the ice cream example the relevant standard deviations are as follows.

Consumption: 0.06579,
Price: 0.00834,
Temperature: 16.422.

The standardized regression coefficients for the two explanatory variables become as shown.

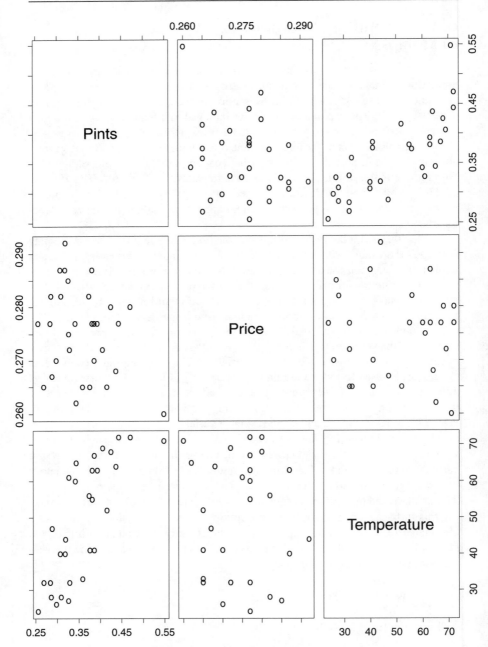

FIG. 6.5.   Scatterplot matrix of variables in ice cream data.

TABLE 6.5
Multiple Regression for the Ice Cream Consumption Data

- The model for the data is

$$\text{consumption} = \beta_0 + \beta_1 \times \text{price} + \beta_2 \times \text{temperature}.$$

- The least-squares estimates of the regression coefficients in the model are as follows.

Parameter	Estimate	SE
β_1	−1.4018	0.9251
β_2	0.0030	0.0005

- The ANOVA table is

Source	SS	DF	MS	F
Regression	0.07943	2	0.03972	23.27
Residual	0.04609	27	0.00171	

Price:

$$\frac{-1.4018 \times 0.00834}{0.06579} = -0.1777. \tag{6.3}$$

Temperature:

$$\frac{0.00303 \times 16.422}{0.06579} = 0.7563. \tag{6.4}$$

A comparison of the standardized coefficients now suggests that temperature is more important in determining consumption.

One further point about the multiple regression model that can usefully be illustrated on this simple example is the effect of entering the explanatory variables into the model in a *different* order. This will become of more relevance in later sections, but simply examining some numerical results from the ice cream data will be helpful for now. The relevant results are shown in Table 6.6. The change produced in a previously estimated regression coefficient when an additional explanatory variable is included in the model results from the correlation between the explanatory variables. If the variables were independent (sometimes the term *orthogonal* is used), their estimated regression coefficients would remain unchanged with the addition of more variables to the model. (As we shall see in Section 6.7, this lack of independence of explanatory variables is also the explanation of the overlapping sums of squares found in unbalanced factorial designs; this topic is introduced in Chapter 4.)

TABLE 6.6
Multiple Regression for the Ice Cream Data: The Effect of Order

Entering Price Followed by Temperature

(a) The first model fitted is

$$\text{consumption} = \beta_0 + \beta_1 \times \text{price}.$$

(b) The parameter estimates in this model are $\hat{\beta}_0 = 0.92$ and $\hat{\beta}_1 = -2.05$.
The multiple correlation coefficient is 0.26 and the ANOVA table is

Source	SS	DF	MS	F
Regression	0.00846	1	0.00846	2.02
Residual	0.1171	28	0.00418	

(c) Temperature is now added to the model; that is, the following new model is fitted:
$$\text{consumption} = \beta_0 + \beta_1 \times \text{price} + \beta_2 \times \text{temperature}.$$

(d) The parameter estimates and ANOVA table are now as in Table 6.5. Note that the estimated regression coefficient for price is now different from the value given in (b) above.

Entering Temperature Followed by Price

(a) The first model fitted is now

$$\text{consumption} = \beta_0 + \beta_1 \times \text{temperature}.$$

(b) The parameter estimates are $\hat{\beta}_0 = 0.21$ and $\hat{\beta}_1 = 0.0031$. The multiple correlation is 0.78 and the ANOVA table is

Source	SS	DF	MS	F
Regression	0.07551	1	0.07551	42.28
Residual	0.05001	28	0.001786	

(c) Price is now added to the model to give the results shown in Table 6.5.

6.3.2. An Example of Multiple Linear Regression in which One of the Explanatory Variables is Categorical

The data shown in Table 6.7 are taken from a study investigating a new method of measuring body composition, and give the body fat percentage, age and sex for 20 normal adults between 23 and 61 years old. The question of interest is 'how are percentage fat, age and sex related?'

The data in Table 6.7 include the categorical variable, sex. Can such an explanatory variable be included in a multiple regression model, and, if so, how? In fact, it is quite legitimate to include a categorical variable such as sex in a multiple

TABLE 6.7
Human Fatness, Age, and Sex

Age	Percentage Fat	Sex
23	9.5	0
23	27.9	1
27	7.8	0
27	17.8	0
39	31.4	1
41	25.9	1
45	27.4	0
49	25.2	1
50	31.1	1
53	34.7	1
53	42.0	1
54	20.0	0
54	29.1	1
56	32.5	1
57	30.3	1
57	21.0	0
58	33.0	1
58	33.8	1
60	33.8	1
61	41.1	1

Note. 1, female; 0, male.

regression model. The distributional assumptions of the model (see Display 6.4) apply only to the response variable. Indeed, the explanatory variables are not strictly considered random variables at all. Consequently the explanatory variables can, in theory at least, be *any* type of variable. However, care is needed in deciding how to incorporate categorical variables with *more* than two categories, as we shall see later in Chapter 10. However, for a two-category variable such as sex there are no real problems, except perhaps in interpretation. Details of the possible regression models for the human fat data are given in Display 6.5.

The results of fitting a multiple regression model to the human fat data, with age, sex, and the interaction of age and sex as explanatory variables, are shown in Table 6.8. The fitted model can be interpreted in the following manner.

1. For men, for whom sex is coded as zero and so sex × age is also zero, the fitted model is

$$\%\text{fat} = 3.47 + 0.36 \times \text{age}. \tag{6.5}$$

Display 6.5
Possible Models that Might be Considered for the Human Fat Data in Table 6.4

- The first model that might be considered is the simple linear regression model relating percentage fat to age:

$$\%\text{fat} = \beta_0 + \beta_1 \times \text{age}.$$

- After such a model has been fitted, a further question of interest might be, Allowing for the effect of age, does a person's sex have any bearing on their percentage of fatness? The appropriate model would be

$$\%\text{fat} = \beta_0 + \beta_1 \times \text{age} + \beta_2 \times \text{sex},$$

where sex is a *dummy variable* that codes men as zero and women as one.
- The model now describes the situation shown in the diagram below, namely two parallel lines with a vertical separation of β_2. Because in this case, the effect of age on fatness is the same for both sexes, or, equivalently, the effect of a person's sex is the same for all ages; the model assumes that there is no *interaction* between age and sex.

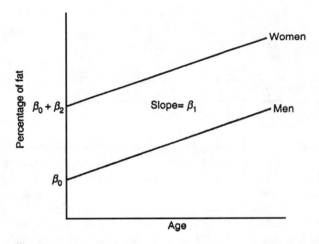

- Suppose now that a model is required that *does* allow for the possibility of an age $\times$ sex interaction effect on the percentage of fatness. Such a model must include a new variable, defined as the product of the variables of age and sex. Therefore, the new model becomes

$$\%\text{fat} = \beta_0 + \beta_1 \times \text{age} + \beta_2 \times \text{sex} + \beta_3 \times \text{age} \times \text{sex}.$$

- To understand this equation better, first consider the percentage of fatness of men; here the values of both sex and sex $\times$ age are zero and the model reduces to

$$\%\text{fat} = \beta_0 + \beta_1 \times \text{age}.$$

(Continued)

Display 6.5
(Continued)

However, for women sex $= 1$ and so sex $\times$ age $=$ age and the model becomes

$$\%\text{fat} + (\beta_0 + \beta_2) + (\beta_1 + \beta_3) \times \text{age}.$$

- Thus the new model allows the lines for males and females to be other than parallel (see diagram below). The parameter β_3 is a measure of the difference between the slopes of the two lines.

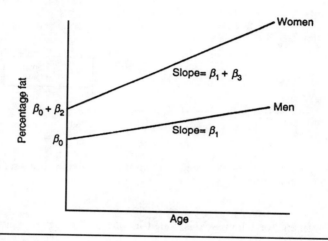

TABLE 6.8
Multiple Regression Results for Human Fatness Data

- The model fitted is

$$\%\text{fat} = \beta_0 + \beta_1 \times \text{age} + \beta_1 \times \text{sex} + \beta_3 \times \text{age} \times \text{sex}.$$

- The least-squares estimates of the regression coefficients in the model are as follows.

Parameter	Estimate	SE
β_1	0.35	0.14
β_2	16.64	8.84
β_3	-0.15	0.19

- The multiple correlation coefficient takes the value 0.8738.
- The regression ANOVA table is

Source	SS	DF	MS	F
Regression	1176.18	3	392.06	17.22
Residual	364.28	16	22.77	

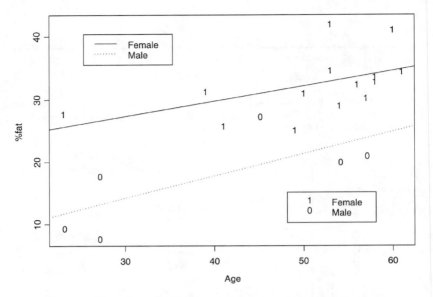

FIG. 6.6. Plot of model for human fat data, which includes a sex × age interaction.

2. For women, for whom sex is coded as one and so sex × age is simply age, the fitted model is

$$\%\text{fat} = 3.47 + 0.36 \times \text{age} + 16.64 - 0.11 \times \text{age}. \tag{6.6}$$

Collecting together terms leads to

$$\%\text{fat} = 20.11 + 0.24 \times \text{age}. \tag{6.7}$$

The fitted model actually represents separate simple linear regressions with different intercepts and different slopes for men and women. Figure 6.6 shows a plot of the fitted model together with the original data.

The interaction effect is clearly relatively small, and it seems that a model including only sex and age might describe the data adequately. In such a model the estimated regression coefficient for sex (11.567) is simply the estimated difference between the percentage fat of men and that of women, which is assumed to be the same at all ages.

(Categorical explanatory variables with more than two categories have to re-coded in terms of a series of binary variables, *dummy variables*, before being used in a regression analysis, but we shall not give such an example until Chapter 10.)

6.3.3. Predicting Crime Rates in the USA: a More Complex Example of Using Multiple Linear Regression

In Table 6.9, data for 47 states of the USA are given. These data will be used to investigate how the crime rate in 1960 depended on the other variables listed. The data originate from the *Uniform Crime Report* of the FBI and other government sources.

Again it is useful to begin our investigation of these data by examining the scatterplot matrix of the data; it is shown in Figure 6.7. The scatterplots involving crime rate (the top row of Figure 6.7) indicate that some, at least, of the explanatory variables are predictive of crime rate. One disturbing feature of the plot is the very strong relationship between police expenditure in 1959 and in 1960. The reasons that such a strongly correlated pair of explanatory variables can cause problems for multiple regression will be taken up in Section 6.6. Here we shall preempt one of the suggestions to be made in that section for dealing with the problem, by simply dropping expenditure in 1960 from consideration.

The results of fitting the multiple regression model to the crime rate data are given in Table 6.10. The global hypothesis that all the regression coefficients are zero is overwhelmingly rejected. The square of the multiple correlation coefficient is 0.75, indicating that the 12 explanatory variables account for 75% of the variability in the crime rates of the 47 states.

The overall test that all regression coefficients in a multiple regression are zero is seldom of great interest. In most applications it will be rejected, because it is unlikely that *all* the explanatory variables chosen for study will be unrelated to the response variable. The investigator is far more likely to be interested in the question of whether some subset of the explanatory variables exists that might be a successful as the full set in explaining the variation in the response variable. If using a particular (small) number of explanatory variables results in a model that fits the data only marginally worse than a much larger set, then a more parsimonious description of the data is achieved (see Chapter 1). How can the most important explanatory variables be identified? Readers might look again at the results for the crime rate data in Table 6.10 and imagine that the answer to this question is relatively straightforward; namely, select those variables for which the corresponding t statistic is significant and drop the remainder. Unfortunately, such a simple approach is only of limited value, because of the relationships between the explanatory variables. For this reason, regression coefficients and their associated standard errors are estimated *conditional* on the other variables in the model. If a variable is removed from a model, the regression coefficients of the remaining variables (and their standard errors) have to be reestimated from a further analysis. (Of course, if the explanatory variables happened to be orthogonal to one another, there would be no problem and the t statistics could be used in selecting the most important explanatory variables. This is, however, of little consequence in most practical applications of multiple regression.)

TABLE 6.9
Crime in the USA: 1960

State	R	Age	S	Ed	Ex0	Ex1	LF	M	N	NW	U1	U2	W	X
1	79.1	151	1	91	58	56	510	950	33	301	108	41	394	261
2	163.5	143	0	113	103	95	583	1012	13	102	96	36	557	194
3	57.8	142	1	89	45	44	533	969	18	219	94	33	318	250
4	196.9	136	0	121	149	141	577	994	157	80	102	39	673	167
5	123.4	141	0	121	109	101	591	985	18	30	91	20	578	174
6	68.2	121	0	110	118	115	547	964	25	44	84	29	689	126
7	96.3	127	1	111	82	79	519	982	4	139	97	38	620	168
8	155.5	131	1	109	115	109	542	969	50	179	79	35	472	206
9	85.6	157	1	90	65	62	553	955	39	286	81	28	421	239
10	70.5	140	0	118	71	68	632	1029	7	15	100	24	526	174
11	167.4	124	0	105	121	116	580	966	101	106	77	35	657	170
12	84.9	134	0	108	75	71	595	972	47	59	83	31	580	172
13	51.1	128	0	113	67	60	624	972	28	10	77	25	507	206
14	66.4	135	0	117	62	61	595	986	22	46	77	27	529	190
15	79.8	152	1	87	57	53	530	986	30	72	92	43	405	264
16	94.6	142	1	88	81	77	497	956	33	321	116	47	427	247
17	53.9	143	0	110	66	63	537	977	10	6	114	35	487	166
18	92.9	135	1	104	123	115	537	978	31	170	89	34	631	165
19	75.0	130	0	116	128	128	536	934	51	24	78	34	627	135
20	122.5	125	0	108	113	105	567	985	78	94	130	58	626	166
21	74.2	126	0	108	74	67	602	984	34	12	102	33	557	195
22	43.9	157	1	89	47	44	512	962	22	423	97	34	288	276
23	121.6	132	0	96	87	83	564	953	43	92	83	32	513	227
24	96.8	131	0	116	78	73	574	1038	7	36	142	42	540	176
25	52.3	130	0	116	63	57	641	984	14	26	70	21	486	196
26	199.3	131	0	121	160	143	631	1071	3	77	102	41	674	152
27	34.2	135	0	109	69	71	540	965	6	4	80	22	564	139
28	121.6	152	0	112	82	76	571	1018	10	79	103	28	537	215
29	104.3	119	0	107	166	157	521	938	168	89	92	36	637	154
30	69.6	166	1	89	58	54	521	973	46	254	72	26	396	237
31	37.3	140	0	93	55	54	535	1045	6	20	135	40	453	200
32	75.4	125	0	109	90	81	586	964	97	82	105	43	617	163
33	107.2	147	1	104	63	64	560	972	23	95	76	24	462	233
34	92.3	126	0	118	97	97	542	990	18	21	102	35	589	166
35	65.3	123	0	102	97	87	526	948	113	76	124	50	572	158
36	127.2	150	0	100	109	98	531	964	9	24	87	38	559	153
37	83.1	177	1	87	58	56	638	974	24	349	76	28	382	254
38	56.6	133	0	104	51	47	599	1024	7	40	99	27	425	225
39	82.6	149	1	88	61	54	515	953	36	165	86	35	395	251
40	115.1	145	1	104	82	74	560	981	96	126	88	31	488	228

(Continued)

TABLE 6.9
(Continued)

41	88.0	148	0	122	72	66	601	998	9	19	84	20	590	144
42	54.2	141	0	109	56	54	523	968	4	2	107	37	489	170
43	82.3	162	1	99	75	70	522	996	40	208	73	27	496	224
44	103.0	136	0	121	95	96	574	1012	29	36	111	37	622	162
45	45.5	139	1	88	46	41	480	968	19	49	135	53	457	249
46	50.8	126	0	104	106	97	599	989	40	24	78	25	593	171
47	84.9	130	0	121	90	91	623	1049	3	22	113	40	588	160

Note. R, crime rate: number of offences known to the police per 1,000,000 population; Age, age distribution: the number of males aged 14–24 years per 1000 of total state population; S, binary variable distinguishing southern states (coded as 1) from the rest; Ed, educational level: mean number of years of schooling × 10 of the population aged ≥25 years; Ex0, police expenditure: per capita expenditure on police protection by state and local government in 1969; Ex1, police expenditure as Ex0, but in 1959; LF, labor force participation rate per 1000 civilian urban males in the age group 14–24 years; M, number of males per 1000 females; N, state population size in hundreds of thousands; NW, number of non-Whites per 1000; U1, unemployment rate of urban males per 1000 in the age group 14–24 years; U2, unemployment rate of urban males per 1000 in the age group 35–39 years; W, wealth as measured by the median value of transferable goods and assets or family income (unit 10 dollars); X, income inequality:the number of families per 1000 earning below 1/2 of the median income.

Because the simple *t* statistics may be misleading when one is trying to choose subsets of explanatory variables, a number of alternative procedures have been suggested; two of these are discussed in the next section.

6.4. SELECTING SUBSETS OF EXPLANATORY VARIABLES

In this section we shall consider two possible approaches to the problem of identifying a subset of explanatory variables likely to be almost as informative as the complete set of variables for predicting the response. The first method is known as *all possible subsets regression*, and the second is known as *automatic model selection*.

6.4.1. All Subsets Regression

Only the advent of modern, high-speed computing and the widespread availability of suitable statistical sofware has made this approach to model selection feasible. We do not intend to provide any details concerning the actual calculations involved, but instead we simply state that the result of the calculations is a table of either all

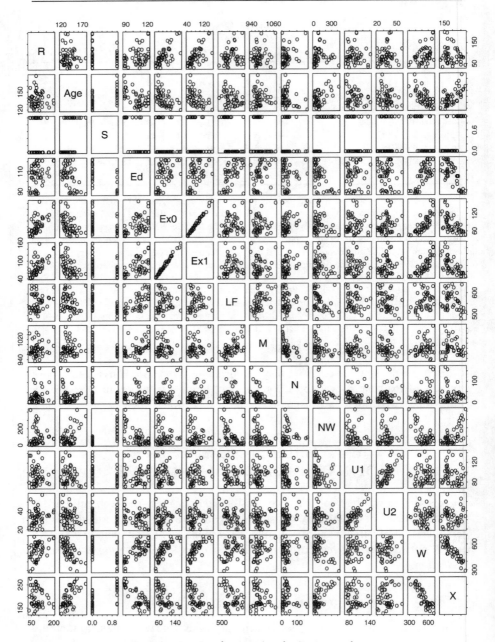

FIG. 6.7. Scatterplot matrix of crime rate data.

TABLE 6.10

Multiple Regression Results for Crime Rate Data (Ex0 not Used)

Parameter	Estimated Coefficients, SEs and t Values			
	Value	SE	t	p
(Intercept)	−739.89	155.55	−4.76	<0.001
Age	1.09	0.43	2.53	0.02
S	−8.16	15.20	−0.54	0.59
Ed	1.63	0.65	2.50	0.02
Ex1	1.03	0.27	3.79	<0.001
LF	0.01	0.15	0.03	0.97
M	0.19	0.21	0.88	0.39
N	−0.016	0.13	−0.13	0.90
NW	−0.002	0.07	−0.03	0.97
U1	−0.62	0.44	−1.39	0.17
U2	1.94	0.87	2.24	0.03
W	0.15	0.11	1.37	0.18
X	0.82	0.24	3.41	0.01

Note. Parameter abbreviations are defined in Table 6.9. Residual standard error, 22.35 on 34 degrees of freedom; Multiple R squared, 0.75; F statistic, 8.64 on 12 and 34 degrees of freedom; the p value is <.001.

possible models, or perhaps a subset of possible models, in which each candidate model is identified by the list of explanantory variables it contains and the values of one or more numerical criteria to use in comparing the various candidates. Such a table is usually organized according to the number of explanatory variables in the candidate model.

For a multiple regression with p explanatory variables, a total of $2^p - 1$ models are possible, because each explanatory variable can be in or out of the model, and the model in which they are all out is excluded. So, for example, for the ice cream data with $p = 2$ explanatory variables, there are three possible models.

Model 1:   temperature,
Model 2:   price,
Model 3:   temperature and price.

In the crime rate example with $p = 12$ explanatory variables, there are $2^{12} - 1 = 4095$ models to consider!

The numerical criterion most often used for assessing candidate models is *Mallows C_k statistic*, which is described in Display 6.6.

Display 6.6

Mallows C_k Statistic

- Mallows C_k statistic is defined as

$$C_k = (\text{RSS}_k / s^2) - (n - 2p),$$

where RSS_k is the residual sum of squares from the multiple regression model with a set of k explanatory variables and s^2 is the estimate of σ^2 obtained from the model that includes all the explanatory variables under consideration.
- If C_k is plotted against k, the subsets of variables worth considering in searching for a parsimonious model are those lying close to the line $C_k = k$.
- In this plot the value of k is (roughly) the contribution to C_k from the variance of the estimated parameters, whereas the remaining $C_k - k$ is (roughly) the contribution from the bias of the model.
- This feature makes the plot a useful device for a broad assessment of the C_k values of a range of models.

TABLE 6.11

Results of all Subsets Regression for Crime Rate Data Using Only
Age, Ed, and Ex1

Model	C_k	Subset Size	Variables in Model
1	9.15	2	Ex1
2	41.10	2	Ed
3	50.14	2	Age
4	3.02	3	Age,Ex1
5	11.14	3	Ed,Ex1
6	42.23	3	Age,Ed
7	4.00	4	Age,Ed,Ex1

Note. The intercept is counted as a term in the model. Abbreviations are defined in Table 6.9.

Now for an example and to begin, we shall examine the all subsets approach on the crime data, but we use only the three explanatory variables, Age, Ed, and Ex1. The results are shown in Table 6.11, and the plot described in Display 6.6 is shown in Figure 6.8. Here only the subset Age and Ex1 appears to be an acceptable alternative to the use of all three variables.

Some of the results from using the all subsets regression technique on the crime rate data with all 12 explanatory variables are given in Table 6.12, and the corresponding plot of C_k values appears in Figure 6.9. Here the plot is too cluttered

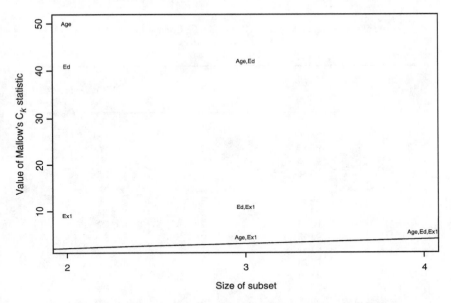

FIG. 6.8.   Plot of C_k against k for all subsets regression, using three explanatory variables in the crime rate data.

to be helpful. However, an inspection of the results from the analysis suggests that the following are two subsets that give a good description of the data.

1. Age, Ed, Ex1, U2, X: $C_6 = 5.65$,
2. Age, Ed, Ex1, W, X: $C_6 = 5.91$.

6.4.2.   Automatic Model Selection

Software packages frequently offer automatic methods of selecting variables for a final regression model from a list of candidate variables. There are three typical approaches: (1) forward selection, (2) backward elimination, and (3) stepwise regression. These methods rely on significance tests known as *partial F tests* to select an explanatory variable for inclusion or deletion from the regression model. The forward selection approach begins with an initial model that contains only a constant term, and it successively adds explanatory variables to the model until the pool of candidate variables remaining contains no variables that, if added to the current model, would contribute information that is statistically important concerning the mean value of the response. The backward elimination method begins with an initial model that contains all the explanatory variables under investigation and then successively removes variables until no variables among those remaining in the model can be eliminated without adversely affecting, in a statistical sense, the predicted value of the mean response. The criterion used for assessing

TABLE 6.12
Some of the Results of All Subsets Regression for Crime Rate Data Using All 12
Explanatory Variables

Model	C_k	Subset Size	Variables in Model
1	33.50	2	Ex1
2	67.89	2	W
3	79.03	2	N
4	80.36	2	Ed
5	88.41	2	M
6	89.80	2	LF
7	90.30	2	X
8	90.38	2	U2
9	93.58	2	S
10	93.61	2	Age
11	20.28	3	Ex1,X
12	23.56	3	Age,Ex1
13	30.06	3	Ex1,M
14	30.89	3	Ex1,NW
15	31.38	3	S,Ex1
16	32.59	3	Ex1,W
17	33.56	3	Ex1,LF
18	34.91	3	Ex1,U2
19	35.46	3	Ex1,U1
20	35.48	3	Ed,Ex1
21	10.88	4	Ed,Ex1,X
22	12.19	4	Ex1,M,X
23	15.95	4	Ex1,LF,X
24	16.82	4	Ex1,W,X
25	18.90	4	Age,Ex1,X
26	20.40	4	Ex1,N,X
27	21.33	4	S,Ex1,X
28	21.36	4	Ex1,NW,X
29	22.19	4	Ex1,U1,X
30	22.27	4	Ex1,U2,X
31	8.40	5	Age,Ed,Ex1,X
32	9.86	5	Ed,Ex1,M,X
33	10.40	5	Ed,Ex1,W,X
34	11.19	5	Ed,Ex1,U2,X
35	12.39	5	Ed,Ex1,LF,X
36	12.47	5	Ed,Ex1,N,X
37	12.65	5	Ed,Ex1,U1,X
38	12.67	5	Age,Ex1,W,X
39	12.85	5	Ed,Ex1,NW,X

(Continued)

TABLE 6.12

(Continued)

Model	C_k	Subset Size	Variables in Model
40	12.87	5	S,Ed,Ex1,X
41	5.65	6	Age,Ed,Ex1,U2,X
42	5.91	6	Age,Ed,Ex1,W,X
43	8.30	6	Age,Ed,Ex1,M,X
44	9.21	6	Age,Ed,Ex1,U1,X
45	9.94	6	Age,Ed,Ex1,NW,X
46	10.07	6	Age,Ed,Ex1,LF,X
47	10.22	6	Age,S,Ed,Ex1,X
48	10.37	6	Age,Ed,Ex1,N,X
49	11.95	6	Ed,Ex1,U1,U2,X
50	14.30	6	Age,Ex1,U2,W,X

Note. Abbreviations are defined in Table 6.9.

whether a variable should be added to an existing model in forward selection, or removed in backward elimination, is based on the change in the residual sum of squares that results from the inclusion or exclusion of a variable. Details of the criterion, and how the forward and backward methods are applied in practice, are given in Display 6.7.

The stepwise regression method of variable selection combines elements of both forward selection and backward elimination. The initial model for stepwise regression is one that contains only a constant term. Subsequent cycles of the approach involve first the possible addition of an explanatory variable to the current model, followed by the possible elimination of one of the variables included earlier, if the presence of new variables has made their contribution to the model no longer significant.

In the best of all possible worlds, the final model selected by applying each of the three procedures outlined above would be the same. Often this does happen, but it is in no way guaranteed. Certainly none of the automatic procedures for selecting subsets of variables are foolproof. They must be used with care, as the following statement from Agresti (1996) makes very clear.

Computerized variable selection procedures should be used with caution. When one considers a large number of terms for potential inclusion in a model, one or two of them that are not really important may look impressive simply due to chance. For instance, when all the true effects are weak, the largest sample effect may substantially

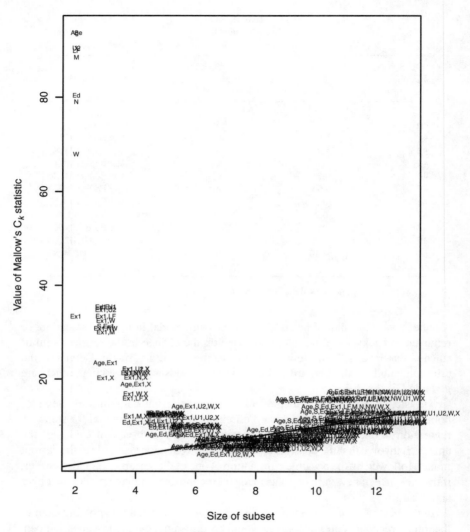

FIG. 6.9. C_k plotted against k for all subsets regression, using 12 explanatory variables in the crime rate data.

overestimate its true effect. In addition, it often makes sense to include variables of special interest in a model and report their estimated effects even if they are not statistically significant at some level.

(See McKay and Campbell, 1982a, 1982b, for some more thoughts on automatic selection methods in regression.)

Display 6.7
Forward Selection and Backward Elimination

- The criterion used for assessing whether a variable should be added to an existing model in forward selection or removed from an existing model in backward elimination is as follows:

$$F = \frac{(\mathrm{SS}_{\mathrm{residual},k} - \mathrm{SS}_{\mathrm{residual},k+1})}{\mathrm{SS}_{\mathrm{residual},\,k+1}/(n-k-2)},$$

where $\mathrm{SS}_{\mathrm{residual},k} - \mathrm{SS}_{\mathrm{residual},k+1}$ is the decrease in the residual sum of squares when a variable is added to an existing k-variable model (forward selection), or the increase in the residual sum of squares when a variable is removed from an existing $(k + 1)$-variable model (backward elimination), and $\mathrm{SS}_{\mathrm{residual},k+1}$ is the residual sum of squares for the model including $k + 1$ explanatory variables.
- The calculated F value is then compared with a preset term know as the F-to-enter (forward selection) or the F-to-remove (backward elimination). In the former, calculated F values greater than the F-to-enter lead to the addition of the candidate variable to the current model. In the latter, a calculated F value less than the F-to-remove leads to discarding the candidate variable from the model.

Application of both forward and stepwise regression to the crime rate data results in selecting the five explanatory variables, Ex1, X, Ed, Age, and U2. Using the backward elimination technique produces these five variables plus W. Table 6.13 shows some of the typical results given by the forward selection procedure implemented in a package such as SPSS.

The chosen subsets largely agree with the selection made by all subsets regression.

It appears that one suitable model for the crime rate data is that which includes Age, Ex1, Ed, U2, and X. The parameter estimates, and so on, for a multiple regression model including only these variables are shown in Table 6.14. Notice that the estimated regression coefficients and their standard errors have changed from the values given in Table 6.10, calculated for a model including the original 12 explanatory variables. The five chosen variables account for 71% of the variabilty of the crime rates (the original 12 explanatory variables accounted for 75%). Perhaps the most curious feature of the final model is that crime rate increases with increasing police expenditure (conditional on the other four explanatory variables). Perhaps this reflects a mechanism whereby as crime increases the police are given more resources?

6.5. REGRESSION DIAGNOSTICS

Having selected a more parsimonious model by using one of the techniques described in the previous section, we still must consider one further important aspect of a regression analysis, and that is to check the assumptions on which the model

TABLE 6.13
Forward Selection Results for Crime Rate Data

Model	Source	SS	DF	MS	F
1	Regression	30586.257	1	30586.257	36.009
	Residual	38223.020	45	849.400	
	Total	68809.277	46		
2	Regression	38187.893	2	19093.947	27.436
	Residual	30621.383	44	695.941	
	Total	68809.227	46		
3	Regression	43886.746	3	14628.915	25.240
	Residual	24922.530	43	579.594	
	Total	68809.277	46		
4	Regression	46124.540	4	11531.135	21.349
	Residual	22684.737	42	540.113	
	Total	68809.277	46		
5	Regression	48500.406	5	9700.081	19.58
	Residual	20308.871	41	495.338	
	Total	68809.277	46		

Model Summary

Model	R	R^2
1 Constant, Ex1	0.667	0.445
2 Constant, Ex1, X	0.745	0.555
3 Constant, Ex1, X, Ed	0.799	0.638
4 Constant, Ex1, X, Ed, Age	0.819	0.670
5 Constant, Ex1, X, Ed, Age, U2	0.840	0.705

Note. No further variables are judged to contribute significantly to the prediction of crime rate. Model summary abbreviations are defined in Table 6.9.

is based. We have, in Section 6.2, already described the use of residuals for this purpose, but in this section we shall go into a little more detail and introduce several other useful regression diagnostics that are now available. These diagnostics are methods for identifying and understanding differences between a model and the data to which it is fitted. Some differences between the data and the model may be the result of isolated observations; one, or a few, observations may be outliers, or may differ in some unexpected way from the rest of the data. Other differences may be systematic; for example, a term may be missing in a linear model.

A number of the most useful regression diagnostics are described in Display 6.8. We shall now examine the use of two of these diagnostics on the final model

TABLE 6.14
Multiple Regression Results for the Final Model Selected for Crime Rate Data

Parameter	Estimate	SE	t	p
Intercept	−528.86	99.62	−5.31	<.001
Age	1.02	0.37	2.76	.009
Ex1	1.30	0.16	8.12	<.001
Ed	2.04	0.50	4.11	<.001
X	0.65	0.16	4.18	<.001
U2	0.99	0.45	2.19	.03

Note. Parameter abbreviations are defined in Table 6.9. $R^2 = 0.705$.

selected for the crime rate data, and Figure 6.10 shows plots of the standard-ized and deletion residuals against fitted values. States 11, 19, and 20 might per-haps be seen as outliers because they fall outside the boundary, $(-2, 2)$. There is also perhaps some evidence that the variance of crime rate is not constant as required. (Exercise 6.6 invites readers to consider more diagnostic plots for these data.)

6.6. MULTICOLLINEARITY

One of the problems that often occurs when multiple regression is used in practice is *multicollinearity*. The term is used to describe situations in which there are moderate to high correlations among some or all of the explanatory variables. Multicollinearity gives rise to a number of difficulties when multiple regression is applied.

1. It severely limits the size of the multiple correlation coefficient R because the explanatory variables are largely attempting to explain much of the same variability in the response variable (see, e.g., Dizney and Gromen, 1967).
2. It makes determining the importance of a given explanatory variable difficult because the effects of the explanatory variables are confounded as a result of their intercorrelations.
3. It increases the variances of the regression coefficients, making the use of the fitted model for prediction less stable. The parameter estimates become unreliable.

Display 6.8
Regression Diagnostics

- To begin we need to introduce the *hat matrix*; this is defined as

$$\mathbf{H} = \mathbf{X}(\mathbf{X'X})^{-1}\mathbf{X'},$$

where $\mathbf{X}$ is the matrix introduced in Display 6.4.
- In a multiple regression the predicted values of the response variable can be written in matrix form as

$$\hat{\mathbf{y}} = \mathbf{Hy},$$

so $\mathbf{H}$ "puts the hats" on $\mathbf{y}$.
- The diagonal elements of $\mathbf{H}$, $h_{ii}, i = 1, \ldots, n$, are such that $0 \le h_{ii} \le 1$, with an average of value p/n. Observations with large values of h_{ii} are said to be leverage points, and it is often informative to produce a plot of h_{ii} against i, an *index plot*, to identify those observations that have high leverage, that is, have most effect on the estimation of parameters.
- The raw residuals introduced in the text are not independent, nor do they have the same variance because $\text{var}(r_i) = \sigma^2(1 - h_{ii})$. Both properties make them less useful than they might be.
- Two alternative residuals are the *standardized residual*, r_i^{std}, and the *deletion residual*, r_i^{del}, defined as follows:

$$r_i^{\text{std}} = \frac{r_i}{\sqrt{[s^2(1 - h_{ii})]}},$$

$$r_i^{\text{del}} = \frac{r_i}{\sqrt{\left[s_{(i)}^2(1 - h_{ii})\right]}},$$

where $s_{(i)}^2$ is the residual mean square estimate of σ^2, after the deletion of observation i.
- The deletion residuals are particularly good in helping to identify outliers.
- A further useful regression diagnostic is *Cook's distance*, defined as

$$D_i = \frac{r_i^2 h_{ii}}{\sqrt{[ps^2(1 - h_{ii})]}}.$$

Cook's distance measures the influence observation i has on estimating all the regression parameters in the model. Values greater than one suggest that the corresponding observation has undue influence on the estimated regression coefficients.
- A full account of regression diagnostics is given in Cook and Weisberg (1982).

Spotting multicollinearity among a set of explanatory variables may not be easy. The obvious course of action is simply to examine the correlations between these variables, but although this *is* often helpful, it is by no means foolproof—more subtle forms of multicollinearity may be missed. An alternative, and generally far more useful, approach is to examine what are know as the *variance inflation*

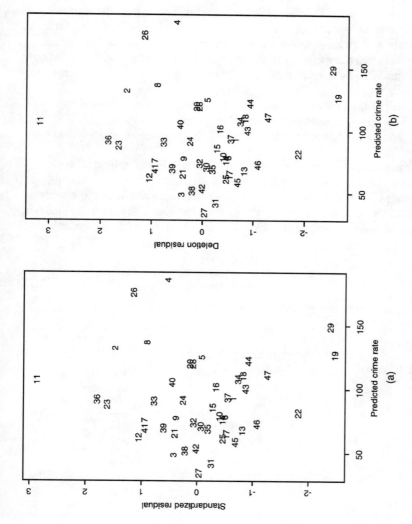

FIG. 6.10. Diagnostic plots for final model selected for crime rate data: (a) stand-ardized residuals against fitted values and (b) deletion residuals against predict-ed values.

factors (VIFs) of the explanatory variables. The variance inflation factor VIF_j, for the jth variable, is given by

$$\text{VIF}_j = 1/(1 - R_j^2),\qquad\qquad(6.8)$$

where R_j^2 is the square of the multiple correlation coefficient from the regression of the jth explanatory variable on the remaining explanatory variables. The variance inflation factor of an explanatory variable indicates the strength of the linear relationship between the variable and the remaining explanatory variables. A rough rule of thumb is that variance inflation factors greater than 10 give some cause for concern.

Returning to the crime rate data, we see that the variance inflation factors of each of the original 13 explanatory variables are shown in Table 6.15. Here it is clear that attempting to use both Ex0 and EX1 in a regression model would have led to problems.

How can multicollinearity be combatted? One way is to combine in some way explanatory variables that are highly correlated—in the crime rate example we could perhaps have taken the mean of Ex0 and Ex1. An alternative is simply to select one of the set of correlated variables (this is the approach used in the analysis of the crime rate data reported previously—only Ex1 was used). Two more complex

TABLE 6.15
VIFs for the Original 13 Explanatory Variables in the
Crime Rate Data

Variable	VIF
Age	2.70
S	4.76
Ed	5.00
Ex0	100.00
Ex1	100.00
LF	3.57
M	3.57
N	2.32
NW	4.17
U1	5.88
U2	5.00
W	10.00
X	8.33

Note. Variable abbreviations are defined in Table 6.9.

possibilities are *regression on principal components* and *ridge regression*, both of which are described in Chatterjee and Price (1991).

6.7. THE EQUIVALENCE OF MULTIPLE REGRESSION AND ANALYSIS OF VARIANCE: THE GENERAL LINEAR MODEL

That (it is hoped) ubiquitous creature, the more observant reader, will have noticed that the models described in this chapter look vaguely similar to those used for analysis of variance, met in Chapters 3 and 4. It is now time to reveal that, in fact, the analysis of variance models and the multiple regression model are *exactly* equivalent.

In Chapter 3, for example, the following model was introduced for a one-way design (see Display 3.2):

$$y_{ij} = \mu + \alpha_i + \epsilon_{ij}. \tag{6.9}$$

Without some constraints on the parameters, the model is overparameterized (see Chapter 3). The constraint usually adopted is to require that

$$\sum_{i=1}^{k} \alpha_i = 0, \tag{6.10}$$

where k is the number of groups. Consequently we can write

$$\alpha_k = -\alpha_1 - \alpha_2 - \cdots - \alpha_{k-1}. \tag{6.11}$$

(Other constraints could be used to deal with the overparameterization problem— see Exercise 6.5.)

How can this be put into an equivalent form to the multiple regression model given in Display 6.4? The answer is provided in Display 6.9, using an example in which there are $k = 3$ groups. Display 6.10 uses the fruit fly data (see Table 3.1) to show how the multiple regression model in Display 6.4 can be used to find the relevant sums of squares in a one-way ANOVA of the data. Notice that the parameter estimates given in Display 6.10 are those to be expected when the model is as specified in Eqs. (6.9) and (6.10), where $\hat{\mu}$ is the overall mean of the number of eggs laid, and the estimates of the α_i are simply the deviations of the corresponding group mean from the overall mean.

Moving on now to the two-way analysis of variance, we find that the simplest way of illustrating the equivalence of the model given in Display 4.1 to the multiple regression model is to use an example in which each factor has only two

Display 6.9

Multiple Regression Model for a One-Way Design with Three Groups

---

- Introduce two variables x_1 and x_2 defined below, to label the group to which an observation belongs.

	Group		
	1	2	3
x_1	1	0	-1
x_2	0	1	-1

- The usual one-way ANOVA model for this situation is

$$y_{ij} = \mu + \alpha_i + \epsilon_{ij},$$

which, allowing for the constraint, $\sum_{i=1}^{3} \alpha_i = 0$, we can now write as

$$y_{ij} = \mu + \alpha_1 x_1 + \alpha_2 x_2 + \epsilon_{ij}.$$

- This is exactly the same form as the multiple regression model in Display 6.4.

Display 6.10

One-Way ANOVA of Fruit Fly Data, Using a Multiple Regression Approach

---

- Define two variables as specified in Display 6.9.
- Regress the number of eggs laid on x_1 and x_2.
- The analysis of variance table from the regression is as follows.

Source	SS	DF	MS	F
Regression	1362	2	681.1	8.7
Residual	5659	72	78.6	

- This is exactly the same as the one-way analysis of variance given in Table 3.2.
- The estimates of the regression coefficients from the analysis are

$$\hat{\mu} = 27.42,$$
$$\hat{\alpha}_1 = -2.16,$$
$$\hat{\alpha}_2 = -3.79.$$

- The estimates of the α_i are simply the differences between each group mean and the grand mean.

---

levels. Display 6.11 gives the details, and Display 6.12 describes a numerical example of performing a two-way ANOVA for a balanced design using multiple regression. Notice that in this case the estimates of the parameters in the multiple regression model do not change when other explanatory variables are added to the model. The explanatory variables for a balanced two-way design are orthogonal (uncorrelated). But now consider what happens when the multiple regression

Display 6.11

Multiple Regression Model for a 2×2 Factorial Design (Factor A at Levels A1 and A2, Factor B at Levels B1 and B2)

- The usual model for a 2×2 design is

$$y_{ijk} = \mu + \alpha_i + \beta_j + \gamma_{ij} + \epsilon_{ijk}.$$

(See Chapter 4.)
- The usual constraints on the parameters introduced to deal with the overparameterized model above are as follows.

$$\sum_{i=1}^{2} \alpha_i = 0,$$

$$\sum_{j=1}^{2} \beta_j = 0,$$

$$\sum_{i=1}^{2} \gamma_{ij} = 0,$$

$$\sum_{j=1}^{2} \gamma_{ij} = 0.$$

- The constraints imply that the parameters in the model satisfy the following equations:

$$\alpha_1 = -\alpha_2,$$
$$\beta_1 = -\beta_2,$$
$$\gamma_{1j} = -\gamma_{2j},$$
$$\gamma_{i1} = -\gamma_{i2}.$$

- The last two equations imply the following:

$$\gamma_{12} = -\gamma_{11}$$
$$\gamma_{21} = -\gamma_{11}$$
$$\gamma_{22} = \gamma_{11}.$$

- In other words, there is only really a single interaction parameter for this design.
- The model for the observations in each of the four cells of the design can now be written explicitly as follows.

	A1	A2
B1	$\mu + \alpha_1 + \beta_1 + \gamma_{11}$	$\mu - \alpha_1 + \beta_1 - \gamma_{11}$
B2	$\mu + \alpha_1 - \beta_1 - \gamma_{11}$	$\mu - \alpha_1 - \beta_1 + \gamma_{11}$

- Now define two variables x_1 and x_2 as follows.

$x_1 = 1$ if first level of A, $x_1 = -1$ if second level of A,

$x_2 = 1$ if first level of B, $x_2 = -1$ if second level of B.

- The original ANOVA model can now be written as

$$y_{ijk} = \mu + \alpha_1 x_1 + \beta_1 x_2 + \gamma_{11} x_3 + \epsilon_{ijk},$$

where $x_3 = x_1 \times x_2$.

Display 6.12
Analysis of a Balanced Two-Way Design by Using Multiple Regression

- Consider the following data, which have four observations in each cell.

	A1	A2
B1	23	22
	25	23
	27	21
	29	21
B2	26	37
	32	38
	30	40
	31	35

- Introduce the three variables x_1, x_2, and x_3 as defined in Display 6.11 and perform a multiple regression, first entering the variables in the order x_1, followed by x_2, followed by x_3. This leads to the following series of results.

    Step 1: x_1 entered. The analysis of variance table from the regression is as follows.

Source	SS	DF	MS
Regression	12.25	1	12.25
Residual	580.75	14	41.48

- The regression sum of squares gives the between levels of A sum of squares that would be obtained in a two-way ANOVA of these data. The estimates of the regression coefficients at this stage are

$$\hat{\mu} = 28.75,$$
$$\hat{\alpha}_1 = -0.875.$$

    Step 2: x_1 and x_2 entered. The analysis of variance for the regression is as follows.

Source	SS	DF	MS
Regression	392.50	2	196.25
Residual	200.50	13	15.42

- The difference in the regression sum of squares between steps 1 and 2, that is, 380.25, gives the sum of squares corresponding to factor B that would be obtained in a conventional analysis of variance of these data. The estimates of the regression coefficients at this stage are

$$\hat{\mu} = 28.75,$$
$$\hat{\alpha}_1 = -0.875,$$
$$\hat{\beta}_1 = -4.875.$$

(Continued)

Display 6.12

(Continued)

Step 3: x_1, x_2, and x_3 entered. The analysis of variance table for this final regression is as follows.

Source	SS	DF	MS
Regression	536.50	3	178.83
Residual	56.50	12	4.71

- The difference in the regression sum of squares between steps 2 and 3, that is, 144.00, gives the sum of squares corresponding to the A × B interaction in an analysis of variance of these data.
- The residual sum of squares in the final table corresponds to the error sum of squares in the usual ANOVA table.
- The estimates of the regression coefficients at this final stage are

$$\hat{\mu} = 28.75,$$

$$\hat{\alpha}_1 = -0.875,$$

$$\hat{\beta}_1 = -4.875,$$

$$\hat{\gamma}_{11} = 3.000.$$

- Note that the estimates of the regression coefficients do *not* change as extra variables are brought into the model.

---

approach is used to carry out an analysis of variance of an unbalanced two-way design, as shown in Display 6.13. In this case, the order in which the explanatory variables enter the multiple regression model *is* of importance; both the sums of squares corresponding to each variable and the parameter estimates depend on which variables have already been included in the model. This highlights the point made previously in Chapter 4 that the analysis of unbalanced designs is not straightforward.

6.8.   SUMMARY

1. Multiple regression is used to assess the relationship between a set of explanatory variables and a continuous response variable.
2. The response variable is assumed to be normally distributed with a mean that is a linear function of the explanatory variables and a variance that is independent of the explanatory variables.
3. The explanatory variables are strictly assumed to be fixed. In practice where this is almost never the case, the results of the multiple regression are to be interpreted conditional on the observed values of these variables.

Display 6.13
Analysis of an Unbalanced Two-Way Design by Using Multiple Regression

- In this case consider the following data.

	A1	A2
B1	23	22
	25	23
	27	21
	29	21
	30	19
	27	23
	23	17
	25	
B2	26	37
	32	38
	30	40
	31	35
		39
		35
		38
		41
		32
		36
		40
		41
		38

- Three variables x_1, x_2, and x_3 are defined as specified in Display 6.11 and used in a multiple regression of these data, with variables entered first in the order x_1, followed by x_2, followed by x_3. This leads to the following series of results.
    Step 1: x_1 entered. The regression ANOVA table is as follows.

Source	SS	DF	MS
Regression	149.63	1	149.63
Residual	1505.87	30	50.19

- The regression sum of squares gives the A sum of squares for an unbalanced design (see Chapter 4).
- The estimates of the regression coefficients at this stage are

$$\hat{\mu} = 29.567,$$
$$\hat{\alpha}_1 = -2.233.$$

Step 2: x_1 entered followed by x_2. The regression ANOVA table is as follows.

Source	SS	DF	MS
Regression	1180.86	2	590.42
Resdual	476.65	29	16.37

(Continued)

Display 6.13

(Continued)

- The increase in the regression sum of squares, that is, 1031.23, is the sum of squares that is due to B, conditional on A already being in the model, that is, B|A as encountered in Chapter 4.
- The estimates of the regression coefficients at this stage are

$$\hat{\mu} = 29.667,$$
$$\hat{\alpha}_1 = -0.341,$$
$$\hat{\beta}_1 = -5.977.$$

Step 3: x_1 and x_2 entered followed by x_3. The regression ANOVA table is as follows.

Source	SS	DF	MS
Regression	1474.25	3	491.42
Residual	181.25	28	6.47

- The increase in the regression sum of squares, that is, 293.39, is the sum of squares that is due to the interaction of A and B, conditional on A and B, that is, AB|A,B.
- The estimates of the regression coefficients at this stage are

$$\hat{\mu} = 28.606,$$
$$\hat{\alpha}_1 = -0.667,$$
$$\hat{\beta}_1 = -5.115$$
$$\hat{\gamma}_{11} = 3.302.$$

- Now enter the variables in the order x_2 followed by x_1 (adding x_3 after x_2 and x_1 will give the same results as step 3 above).

  Step 1: x_2 entered. The regression ANOVA table is as follows.

Source	SS	DF	MS
Regression	1177.70	1	1177.70
Residual	477.80	30	15.93

- The regression sum of squares is that for B for an unbalanced design.
- The estimates of the regression coefficients at this stage are

$$\hat{\mu} = 29.745,$$
$$\hat{\beta}_1 = -6.078.$$

Step 2: x_2 entered followed by x_1. The regression ANOVA table is as follows.

Source	SS	DF	MS
Regression	1180.85	2	590.42
Residual	474.65	29	16.37

(Continued)

Display 6.13
(Continued)

- The estimates of the regression coefficients at this stage are

$$\hat{\mu} = 29.667,$$
$$\hat{\alpha}_1 = -0.341,$$
$$\hat{\beta}_1 = -5.977.$$

- Note how the regression estimates for a variable alter, depending on which stage the variable is entered into the model.

4. It may be possible to find a more parsimonious model for the data, that is, one with fewer explanatory variables, by using all subsets regression or one of the "stepping" methods. Care is required when the latter is used as implemented in a statistical package.
5. An extremely important aspect of a regression analysis is the inspection of a number of regression diagnostics in a bid to identify any departures from assumptions, outliers, and so on.
6. The multiple linear regression model and the analysis of variance models described in earlier chapters are equivalent.

COMPUTER HINTS

SPSS

To conduct a multiple regression analysis with one set of explanatory variables, use the following basic steps.

1. Click **Statistics**, click **Regression**, and click **Linear**.
2. Click the dependent variable from the variable list and move it to the Dependent box.
3. Hold down the ctrl key, and click on the explanatory variables and move them to the Independent box.
4. Click **Statistics**, and click **Descriptives**.

S-PLUS

In S-PLUS, regression can be used by means of the Statistics menu as follows.

1. Click **Statistics**, click **Regression**, and click **Linear** to get the Multiple Regression dialog box.
2. Select dependent variable and explanatory variables, or define the regression of interest in the Formula box.
3. Click on the **Plot** tag to select residual plots, and so on.

Multiple regression can also be applied by using the command language with the *lm* function with a formula specifying what is to be regressed on what. For example, if the crime in the USA data in the text are stored in a data frame, *crime* with variable names as in the text, multiple regression could be applied by using

$$lm(R \sim Age + S + Ed + Ex1 + LF + M + N + NW + U1 + U2 + W + X).$$

After fitting a model, S-PLUS has extensive facilities for finding many types of regression diagnostics and plotting them.

All subsets regression is available in S-PLUS by using the *leaps* function.

EXERCISES

6.1. Explore other possible models for the vocabulary data. Possibilities are to include a quadratic age term or to model the log of vocabulary size.

6.2. Examine the four data sets shown in Table 6.16. Also show that the estimates of the slope and intercept parameters in a simple linear regression are the

TABLE 6.16

Four Hypothetical Data Sets, Each Containing 11 Observations for Two Variables

				Data Set		
	1–3	*1*	*2*	*3*	*4*	*4*
				Variable		
Observation	*x*	*y*	*y*	*y*	*x*	*y*
1	10.0	8.04	9.14	7.46	8.0	6.58
2	8.0	6.95	8.14	6.77	8.0	5.76
3	13.0	7.58	8.74	12.74	8.0	7.71
4	9.0	8.81	8.77	7.11	8.0	8.84
5	11.0	8.33	9.26	7.81	8.0	8.47
6	14.0	9.96	8.10	8.84	8.0	7.04
7	6.0	7.24	6.13	6.08	8.0	5.25
8	4.0	4.26	3.10	5.39	19.0	12.50
9	12.0	10.84	9.13	8.15	8.0	5.56
10	7.0	4.82	7.26	6.42	8.0	7.91
11	5.0	5.68	4.74	5.73	8.0	6.89

TABLE 6.17
Memory Retention

t	p
1	0.84
5	0.71
15	0.61
30	0.56
60	0.54
120	0.47
240	0.45
480	0.38
720	0.36
1440	0.26
2880	0.20
5760	0.16
10080	0.08

TABLE 6.18
Marriage and Divorce Rates per 1000 per Year
for 14 Countries

Marriage Rate	Divorce Rate
5.6	2.0
6.0	3.0
5.1	2.9
5.0	1.9
6.7	2.0
6.3	2.4
5.4	0.4
6.1	1.9
4.9	2.2
6.8	1.3
5.2	2.2
6.8	2.0
6.1	2.9
9.7	4.8

TABLE 6.19
Quality of Children's Testimonies

Age	Gender	Location	Coherence	Maturity	Delay	Prosecute	Quality
5–6	Male	3	3.81	3.62	45	No	34.11
5–6	Female	2	1.63	1.61	27	Yes	36.59
5–6	Male	1	3.54	3.63	102	No	37.23
5–6	Female	2	4.21	4.11	39	No	39.65
5–6	Male	3	3.30	3.12	41	No	42.07
5–6	Female	3	2.32	2.13	70	Yes	44.91
5–6	Female	4	4.51	4.31	72	No	45.23
5–6	Female	2	3.18	3.08	41	No	47.53
8–9	Male	3	3.02	3.00	71	No	54.64
8–9	Female	2	2.77	2.71	56	Yes	57.87
8–9	Male	3	3.35	3.07	88	Yes	57.07
5–6	Male	1	2.66	2.72	13	No	45.81
5–6	Female	3	4.70	4.98	29	No	49.38
5–6	Male	2	4.31	4.21	39	Yes	49.53
8–9	Female	2	2.16	2.91	10	No	67.08
8–9	Male	4	1.89	1.87	15	Yes	83.15
8–9	Female	2	1.94	1.99	46	Yes	80.67
8–9	Female	3	2.86	2.93	57	No	78.47
5–6	Female	4	3.11	3.01	26	Yes	77.59
8–9	Male	2	2.90	2.87	14	No	76.28
8–9	Male	4	2.41	2.38	45	No	59.64
8–9	Male	4	2.32	2.33	19	Yes	68.44
5–6	Male	4	2.78	2.79	9	Yes	65.07

same for each set. Find the value of the multiple correlation coefficient for each data set. (This example illustrates the dangers of blindly fitting a regression model without the use of some type of regression diagnostic.)

6.3. Plot some suitable diagnostic graphics for the four data sets in Table 6.16.

6.4. The data shown in Table 6.17 give the average percentage of memory retention, p, measured against passing time, t (minutes). The measurements were taken five times during the first hour after subjects memorized a list of disconnected items, and then at various times up to a week later. Plot the data (after a suitable transformation if necessary) and investigate the relationship between retention and time by using a suitable regression model.

6.5. As mentioned in Chapter 3, ANOVA models are usually presented in over-parameterized form. For example, in a one-way analysis of variance, the constraint

$\sum_{i=1}^{k} \alpha_i = 0$ is often introduced to overcome the problem. However, the overparameterization in the one-way ANOVA can also be dealt with by setting one of the α_i equal to zero. Carry out a multiple regression of the fruit fly data that is equivalent to the ANOVA model with $\alpha_3 = 0$. In terms of group means, what are the parameter estimates in this case?

6.6. For the final model used for the crime rate data, examine plots of the diagonal elements of the hat matrix against observation number, and produce a similar index plot for the values of Cook's distance statistic. Do these plots suggest any amendments to the analysis?

6.7. Table 6.18 shows marriage and divorce rates (per 1000 populations per year) for 14 countries. Derive the linear regression equation of divorce rate on marriage rate and show the fitted line on a scatterplot of the data. On the basis of the regression line, predict the divorce rate for a country with a marriage rate of 8 per 1000 and also for a country with a marriage rate of 14 per 1000. How much conviction do you have in each prediction?

6.8. The data shown in Table 6.19 are based on those collected in a study of the quality of statements elicited from young children. The variables are statement quality; child's gender, age, and maturity; how coherently the child gave evidence; the delay between witnessing the incident and recounting it; the location of the interview (the child's home, school, a formal interviewing room, or an interview room specially constructed for children); and whether or not the case proceeded to prosecution. Carry out a complete regression analysis on these data to see how statement quality depends on the other variables, including selecting the best subset of explanatory variables and examining residuals and other regression diagnostics. Pay careful attention to how the categorical explanatory variables with more than two categories are coded.

7

Analysis of Longitudinal Data

7.1. INTRODUCTION

Longitudinal data were introduced as a special case of repeated measures in Chapter 5. Such data arise when subjects are measured on the same variable (or, in some cases, variables), on several different occasions. Because the repeated measures arise, in this case, solely from the passing of time, there is no possibility of randomizing the "occasions," and it is this that essentially differentiates longitudinal data from other repeated measure situations arising in psychology, where the subjects are observed under different conditions, or combinations of conditions, which are usually given in different orders to different subjects. The special structure of longitudinal data makes the sphericity condition described in Chapter 5, and methods that depend on it for their validity, very difficult to justify. With longitudinal data it is very unlikely that the measurements taken close to one another in time will have the same correlation as measurements made at more widely spaced time intervals.

The analysis of longitudinal data has become something of a growth industry in statistics, largely because of its increasing importance in clinical trials (see Everitt and Pickles, 2000). Here we shall cover the area only briefly, concentrating on one very simple approach and one more complex, regression-based modeling procedure.

However, first the question of why the observations for each occasion should not be separately analyzed has to be addressed. When two groups of subjects are being compared as with the salsolinol data in Chapter 5 (see Table 5.2), this occasion-by-occasion approach would involve a series of t tests. When more than two groups are present, a series of one-way analyses of variance would be required. (Alternatively, some distribution-free equivalent might be used; see Chapter 8.) The procedure is straightforward but has a number of serious flaws and weaknesses. The first is that the series of tests performed are not independent of one another, making interpretation difficult. For example, a succession of marginally significant group mean differences might be collectively compelling if the repeated measurements are only weakly correlated, but much less convincing if there is a pattern of strong correlations. The occasion-by-occasion procedure also assumes that each repeated measurement is of separate interest in its own right. This is unlikely in general, because the real concern is likely to involve something more global; in particular, the series of separate tests provide little information about the longitudinal development of the mean response profiles. In addition, the separate significance tests do not given an overall answer to whether or not there is a group difference and, of particular importance, do not provide a useful estimate of the overall "treatment" effect.

7.2. RESPONSE FEATURE ANALYSIS: THE USE OF SUMMARY MEASURES

A relatively straightforward approach to the analysis of longitudinal data is to first transform the T (say) repeated measurements for each subject into a single number considered to capture some important aspect of a subject's response profile; that is, for each subject, calculate s given by

$$s = f(x_1, x_2, \ldots, x_T), \tag{7.1}$$

where $x_1, x_2, \ldots, x_T$ are the repeated measures for the subject and f represents the chosen summary function. The chosen summary measure s has to be decided on before the analysis of the data begins and should, of course, be relevant to the particular questions that are of interest in the study. Commonly used summary measures are (1) overall mean, (2) maximum (minimum) value, (3) time to maximum (minimum) response, (4) slope of regression line of response on time, and (5) time to reach a particular value (e.g., a fixed percentage of baseline).

Having identified a suitable summary measure, the analysis of differences in the levels of the between subject factor(s) is reduced to a simple t test (two groups) or an analysis of variance (more than two groups or more than a single between subjects factor. Alternatively, the distribution-free equivalents might be used if there is any evidence of a departure from normality in the chosen summary measure.)

As a first example of the use of the summary measure approach to longitudinal data, it will be applied to the salsolinol data given in Chapter 5. (Because the data are clearly very skewed, we shall analyze the log-transformed observations.) We first need to ask, What is a suitable summary measure for these data? Here the difference in average excretion level is likely to be one question of interest, so perhaps the most obvious summary measure to use is simply the average response of each subject over the four measurement occasions. Table 7.1 gives the values of this summary measure for each subject and also the result of constructing a confidence interval for the difference in the two groups. The confidence interval contains the value zero, and so, strictly, no claim can be made that the data give any evidence of a difference between the average excretion levels of those people who are moderately dependent and those who are severely dependent on alcohol. However, when a more pragmatic view is taken of the lack of symmetry of the confidence interval around zero, it does appear that subjects who are severely dependent on alcohol may tend to have a somewhat higher salsolinol excretion level.

Now let us examine a slightly more complex example. The data given in Table 7.2 result from an experiment in which three groups of rats were put on different diets, and after a settling in period their bodyweights (in grams) were recorded weekly for nine weeks. Bodyweight was also recorded once before the diets began. Some observations are missing (indicted by NA in Table 7.2) because of the failure to record a bodyweight on particular occasions as intended. Before the response feature approach is applied to these data, two possible complications have to be considered.

1. What should be done about the prediet bodyweight observation taken on day 1?
2. How should the missing observations be dealt with?

We shall leave aside the first of these questions for the moment and concentrate on only the nine postdiet recordings. The simplest answer to the missing values problem is just to remove rats with such values from the analysis. This would leave 11 of the original 16 rats for analysis. This approach may be simple but it is not good! Using it in this case would mean an almost 33% reduction in sample size—clearly very unsatisfactory.

An alternative to simply removing rats with any missing values is to calculate the chosen summary measure from the *available* measurements on a rat. In this way, rats with different numbers of repeated measurements can all contribute to the analysis. Adopting this approach, and again using the mean as the chosen summary measures, leads to the values in Table 7.3 on which to base the analysis. The results of a one-way analysis of variance of these measures, and the Bonferonni

TABLE 7.1
Summary Measure Results for Logged Salsolinol Data
(log to base 10 used)

Average Response for Each Subject		
Group	Subject	Average Response
Group 1 (moderate)	1	0.06
	2	0.19
	3	0.19
	4	−0.02
	5	−0.08
	6	0.17
Group 2 (severe)	1	−0.05
	2	0.28
	3	0.52
	4	0.20
	5	0.05
	6	0.41
	7	0.37
	8	0.06

Means and Standard Deviations			
Group	Mean	SD	n
Moderate	0.0858	0.1186	6
Severe	0.2301	0.1980	8

Note. Log to base 10 is used. The 95% confidence interval for the difference between the two groups is

$$(0.2301 - 0.0858) \pm \times s \times (1/6 + 1/8)^{1/2}$$

when s^2 is the assumed common variance in the two groups and is given by

$$s^2 = \frac{5 \times 0.1186^2 + 7 \times 0.1980^2}{8 + 6 - 2} = 0.2873.$$

These calculations lead to the interval (−0.0552, 0.3438).

multiple comparisons tests, are given in Table 7.4. (The Scheffé procedure gives almost identical results for these data.) Clearly the mean bodyweight of Group 1 differs from the mean bodyweight of Groups 2 and 3, but the means for the latter do not differ.

(It should be noted here that when the number of available repeated measures differs from subject to subject in a longitudinal study, the calculated summary

TABLE 7.2
Bodyweights of Rats (grams)

Group	Day									
	1	8	15	22	29	36	43	50	57	64
1	240	250	255	260	262	258	266	265	272	278
1	225	230	230	NA	240	240	243	238	247	245
1	245	250	250	255	262	265	267	264	268	269
1	260	255	255	265	265	268	270	274	273	275
1	255	260	255	270	270	273	NA	276	278	280
1	260	265	270	275	275	277	278	284	279	281
1	275	275	260	270	NA	274	276	282	NA	284
1	245	255	260	268	270	265	265	273	274	278
2	410	415	425	428	438	443	442	456	468	478
2	405	420	430	440	448	460	458	475	484	496
2	445	445	450	NA	455	455	451	462	466	472
2	555	560	565	580	590	597	595	612	618	628
3	470	465	475	485	487	493	493	507	518	525
3	535	525	530	533	535	540	525	543	544	559
3	520	525	530	540	545	546	538	553	555	548
3	510	510	520	NA	530	538	535	550	553	NA

Note. NA, missing value: 1, Diet 1; 2, Diet 2; 3, Diet 3. Diets begin after the measurement of bodyweight on day 1.

measures are likely to have differing precisions. The analysis should then, ideally, take this into account by using some form of weighting. This was not attempted in the rat data example because it involves techniques outside the scope of this text. However, for those readers courageous enough to tackle a little mathematics, a description of a statistically more satisfactory procedure is given in Gornbein, Lazaro, and Little, 1992.)

Having addressed the missing value problem, we now need to consider if and how to incorporate the prediet bodyweight recording (known generally as a *baseline* measurement) into an analysis. Such a pretreatment value can be used in association with the response feature approach in a number of ways. For example, if the average response over time is the chosen summary (as in the two examples above), there are three possible methods of analysis (Frison and Pocock, 1992).

1. POST—an analysis that ignores the baseline values available and analyzes only the mean of the postdiet responses (the analysis reported in Table 7.4).

TABLE 7.3
Means of Available Posttreatment Observations
for Rat Data

Group	Rat	Avg. Bodyweight
Group 1	1	262.9
	2	239.2
	3	261.1
	4	266.9
	5	270.6
	6	276.0
	7	272.7
	8	267.8
Group 2	1	444.1
	2	457.4
	3	456.9
	4	593.9
Group 3	1	495.4
	2	537.7
	3	542.7
	4	534.7

TABLE 7.4
ANOVA of Means Given in Table 7.3

Source	DF	SS	MS	F	p
Group	2	239475	119738	89.6	<.001
Error	13	17375	1337		

Bonferroni Multiple Comparisons

Comparison	Estimate	SE	Lower Bound	Upper Bound
g1–g2	−223.0	22.4	−285	−12[a]
g1–g3	−263.0	22.4	−324	−201[a]
g2–g3	−39.5	25.9	−111	31.4

[a] Intervals excluding zero.

2. CHANGE—an analysis that involves the differences of the mean postdiet weights and the prediet value.
3. ANCOVA—here between subject variation in the prediet measurement is taken into account by using the prediet recording as a covariate in a linear model for the comparison of postdiet means.

Each method represents a different way of adjusting the observed difference at outcome by using the prediet difference. POST, for example, relies on the fact that when groups are formed by randomization, then in the absence of a difference between diets, the expected value of the mean difference at outcome is zero. Hence, the factor by which the observed outcome requires correction in order to judge the diet effect is also zero.

The CHANGE approach corresponds to the assumption that the difference at outcome, in the absence of a diet effect, is expected to be equal to the differences in the means of the prediet values. That such an assumption is generally false is well documented—see, for example, Senn (1994a, 1994b). The difficulty primarily involves the *regression to the mean* phenomenon. This refers to the process that occurs as transient components of an initial measurement are dissipated over time. Selection of high-scoring individuals for entry into a study, for example, necessarily also selects for individuals with high values of any transient component that might contribute to that score. Remeasurement during the study will tend to show a declining mean value for such groups. Consequently, groups that initially differ through the existence of transient phenomena, such as some forms of measurement error, will show a tendency to have converged on remeasurement. Randomization ensures only that treatment groups are similar in terms of expected values and so may actually differ not just in transient phenomena but also in more permanent components of the observed scores. Thus, although the description of transient components may bring about the regression to the mean phenomena such as those previously described, the extent of regression and the mean value to which separate groups are regressing need not be expected to be the same. Analysis of covariance (ANCOVA) allows for such differences. The use of ANCOVA allows for some more general system of predicting what the outcome difference would have been in the absence of any diet effect, as a function of the mean difference prediet.

Frison and Pocock (1992) compare the three approaches and show that an analysis of covariance is more powerful than both the analysis of change scores and analysis of posttreatment means only. (Using the mean of several baseline values, if available, makes the analysis of covariance even more efficient, if there are moderate correlations between the repeated mean.) The differences between the three approaches can be illustrated by comparing power curves calculated by using the results given in Frison and Pocock (1992). Figures 7.1, 7.2, and 7.3 show some examples for the situation with two treatment groups, a pretreatment measure, nine posttreatment measures, and varying degrees of correlation between the repeated

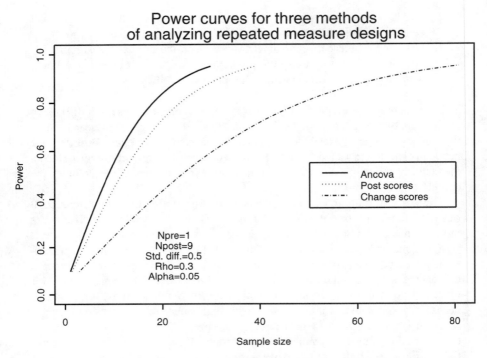

Power curves for three methods
of analyzing repeated measure designs

FIG. 7.1. Power curves comparing POST, CHANGE, and ANCOVA.

observations (the correlations between pairs of repeated measures are assumed equal in the calculation of these power curves). From these curves, it can be seen that the sample size needed to achieve a particular power for detecting a standardized treatment difference of 0.5 is always *lower* with an ANCOVA. When the correlation between pairs of repeated measures is small (0.3), CHANGE is worse than simply ignoring the pretreatment value available and simply using POST. As the correlation increases, CHANGE approaches ANCOVA in power, with both being considerably better than POST.

The results of applying each of CHANGE and ANCOVA to the rat data are given in Table 7.5. (The POST results were given previously in Table 7.4). Here both POST and CHANGE suggest the presence of group differences, but ANCOVA finds the group difference not significant at the 5% level, suggesting that the observed postdiet differences are accountable for by the prediet differences. It is clear that the three groups of rats in this experiment have very different initial bodyweights, and the warnings about applying an analysis of covariance in such cases spelled out in Chapter 3 have to be kept in mind. Nevertheless, it might be argued that the postdiet differences in bodyweight between the three groups are

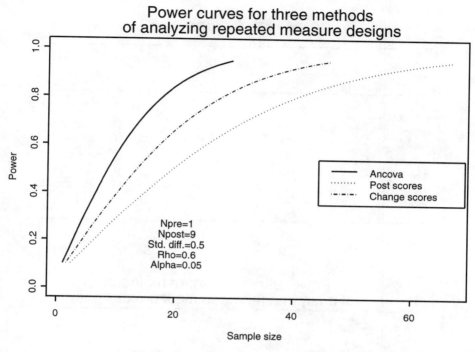

FIG. 7.2. Power curves comparing POST, CHANGE, and ANCOVA.

not the results of the different diets. A telling question that might be asked here is, why were the three diets were tried on animals, some of which were markedly lighter than the rest?

The response feature method has a number of advantages for the analysis of longitudinal data. Three given by Matthews (1993) are as follows.

1. An appropriate choice of summary measure ensures that the analysis is focused on relevant and interpretable aspects of the data.
2. The method is statistically respectable.
3. To some extent, missing and irregularly spaced observations can be accommodated.

A disadvantage of the response feature approach is that it says little about how the observations evolve over time, and whether different groups of subjects behave in a similar way as time progresses. When such issues are of interest, as they generally are, an alternative is needed, such as the more formal modeling procedures that are described in the next section.

TABLE 7.5
CHANGE and ANCOVA for the Rat Data

Source	DF	SS	MS	F	p
CHANGE					
Group	2	1098	549.1	4.30	.037
Error	13	1659	127.6		
ANCOVA					
Pretreatment	1	254389.0	254389.0	1972.01	<.0001
Group	2	912	456.0	3.53	.062
Error	12	1550	129		

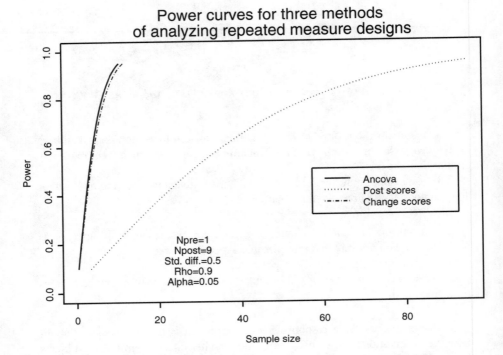

FIG. 7.3. Power curves comparing POST, CHANGE, and ANCOVA.

7.3. RANDOM EFFECT MODELS
FOR LONGITUDINAL DATA

A detailed analysis of longitudinal data requires consideration of models that represent both the level and the slope of a group's profile of repeated measurements, and also account adequately for the observed pattern of dependences in those measurements. Such models are now widely available, but confusingly they are often referred to by different names, including, for example, *random growth curve models*, *multilevel models*, *random effects models*, and *hierarchical models*. Such models have witnessed a huge increase in interest in recent years, and there is now a considerable literature surrounding them. Here we shall only try to give the flavor of the possibilities, by examining one relatively straightforward example. Fuller accounts of this increasingly important area of statistics are available in Brown and Prescott (1999) and Everitt and Pickles (2000).

7.3.1. The Treatment of Alzheimer's Disease

The data in Table 7.6 arise from an investigation of the use of lecithin, a precursor of choline, in the treatment of Alzheimer's disease. Traditionally, it has been assumed that this condition involves an inevitable and progressive deterioration in all aspects of intellect, self-care, and personality. Recent work suggests that the disease involves pathological changes in the central cholinergic system, which it might be possible to remedy by long-term dietary enrichment with lecithin. In particular, the treatment might slow down or perhaps even halt the memory impairment associated with the condition. Patients suffering from Alzheimer's disease were randomly allocated to receive either lecithin or placebo for a 6-month period. A cognitive test score giving the number of words recalled from a previously standard list was recorded monthly for 5 months.

As with most data sets, it is important to graph the data in Table 7.6 in some informative way prior to a more formal analysis. According to Diggle, Liang, and Zeger (1994), there is no single prescription for making effective graphical displays of longitudinal data, although they do offer the following simple guidelines.

- show as much of the relevant raw data as possible rather than only data summaries,
- highlight aggregate patterns of potential scientific interest,
- identify both cross-sectional and longitudinal patterns, and
- make easy the identification of unusual individuals or unusual observations.

Three graphical displays of the data in Table 7.6 are shown in Figures 7.4, 7.5, and 7.6. In the first, the raw data for each patient are plotted, and they are labeled with the group to which they have been assigned. There is considerable variability

TABLE 7.6
The Treatment of Alzheimer's Disease

Group	Visit				
	1	2	3	4	5
1	20	15	14	13	13
1	14	12	12	10	10
1	7	5	5	6	5
1	6	10	9	8	7
1	9	7	9	5	4
1	9	9	9	11	8
1	7	3	7	6	5
1	18	17	16	14	12
1	6	9	9	9	9
1	10	15	12	12	11
1	5	9	7	3	5
1	11	11	8	8	9
1	10	2	9	3	5
1	17	12	14	10	9
1	16	15	12	7	9
1	7	10	4	7	5
1	5	0	5	0	0
1	16	7	7	6	4
1	2	1	1	2	2
1	7	11	7	5	8
1	9	16	14	10	6
1	2	5	6	7	6
1	7	3	5	5	5
1	19	13	14	12	10
1	7	5	8	8	6
2	9	12	16	17	18
2	6	7	10	15	16
2	13	18	14	21	21
2	9	10	12	14	15
2	6	7	8	9	12
2	11	11	12	14	16
2	7	10	11	12	14
2	8	18	19	19	22
2	3	3	3	7	8
2	4	10	11	17	18
2	11	10	10	15	16
2	1	3	2	4	5
2	6	7	7	9	10
2	0	3	3	4	6
2	18	18	19	22	22
2	15	15	15	18	19
2	10	14	16	17	19
2	6	6	7	9	10
2	9	9	13	16	20
2	4	3	4	7	9
2	4	13	13	16	19
2	10	11	13	17	21

Note. 1, placebo; 2, lecithin groups.

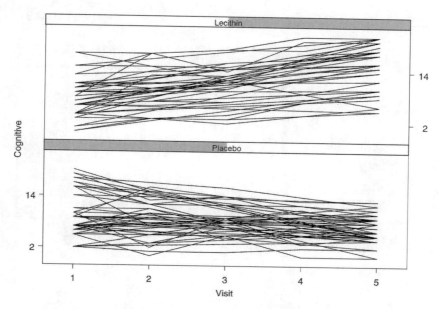

FIG. 7.4. Plot of individual patient profiles for the lecithin trial data in Table 7.6.

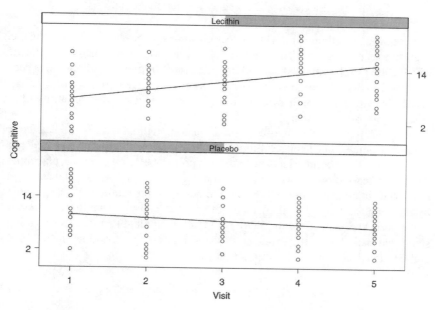

FIG. 7.5. Plot of individual patient data from the lecithin trial for each treatment group, showing fitted linear regressions of cognitive score on visit.

221

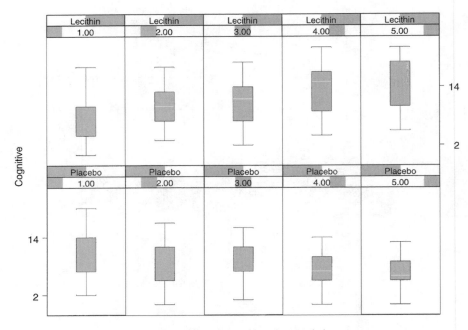

FIG. 7.6. Box plots of lecithin trial data.

in these profiles, although patients on lecithin appear to have some increase in their cognitive score over time. This observation appears to be confirmed by Figure 7.5, which shows each group's data, and the fitted linear regression of cognitive score on visit. Finally the box plots given in Figure 7.6 for each group and each visit give little evidence of outliers, or of worrying differences in the variability of the observations on each visit.

A further graphic that is often useful for longitudinal data is the scatterplot matrix introduced in Chapter 2. Figure 7.7 shows such a plot for the repeated measurements in the lecithin trial data, with each point labeled by group (0, placebo; 1, lecithin). Clearly, some pairs of measurements are more highly related than others.

Now let us consider some more formal models for this example. We might begin by ignoring the longitudinal character of the data and use a multiple regression model as described in Chapter 6. We could, for example, fit a model for cognitive score that includes, as the explanatory variables, Visit (with values 1, 2, 3, 4, 5) and Group (0, placebo; 1, lecithin). That is,

$$y_{ijk} = \beta_0 + \beta_1 \text{Visit}_j + \beta_2 \text{Group}_i + \epsilon_{ijk}, \qquad (7.2)$$

where y_{ijk} is the cognitive score for subject k on visit j, in group i, and the ϵ_{ijk} are error terms assumed normally distributed with mean zero and variance σ^2. Note

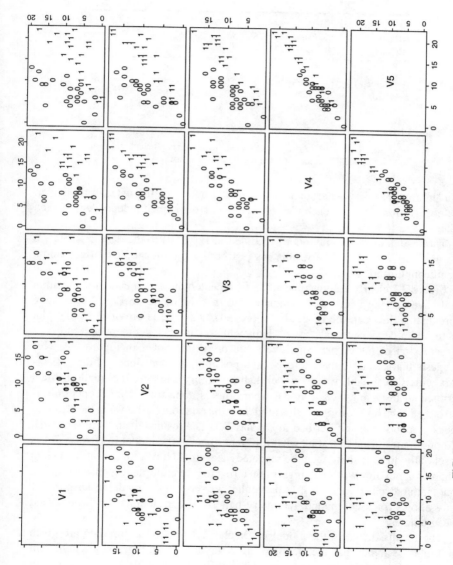

FIG. 7.7. Scatterplot matrix of the cognitive scores on the five visits of the lecithin trial, showing treatment group (0, placebo; 1, lecithin).

223

TABLE 7.7
Results of Fitting the Regression Model Described in Eq. (7.2)
to the Data in Table 7.6

Parameter	Estimate	SE	t	p
β_0(Intercept)	6.93	0.802	8.640	<.001
β_1(Visit)	0.494	0.224	2.199	.029
β_2(Group)	3.056	0.636	4.803	<.001

Note. $R^2 = 0.107$; F for testing that all regression coefficients are zero
is 14 on 2 and 232 degrees of freedom; p value is <.001.

that this model takes no account of the real structure of the data; in particular, the observations made at each visit on a subject are assumed independent of each other.

The results of fitting the model specified in Eq. (7.2) to the data in Table 7.6 are given in Table 7.7. The results suggest both a significant Visit effect, and a significant Group effect.

Figure 7.7 clearly demonstrates that the repeated measurements in the lecithin trial have varying degrees of association so that the model in Eq. (7.2) is quite unrealistic. How can the model be improved? One clue is provided by the plot of the confidence intervals for each patient's estimated intercept and slope in a regression of his or her cognitive score on Visit, shown in Figure 7.8. This suggests that certainly the intercepts and possibly also the slopes of the patients vary widely. A suitable way to model such variability is to introduce random effect terms, and in Display 7.1 a model where this is done for the intercepts only is described. The implications of this model for the relationship between the repeated measurements is also described in Display 7.1. The results given there show that although introduced in a somewhat different manner, the model in Display 7.1 is essentially nothing more than the repeated measure ANOVA model introduced in Chapter 5, but now written in regression terms for the multiple response variables, that is, the repeated measures. In particular, we see that the model in Display 7.1 only allows for a compound symmetry pattern of correlations for the repeated mesurements.

The results of fitting the model in Display 7.1 to the lecithin data are given in Table 7.8. Note that the standard error for the Group regression coefficient is greater in the random effects model than in the multiple regression model (see Table 7.7), and the reverse is the case for the regression coefficient of visit. This arises because the model used to obtain the results in Table 7.2,

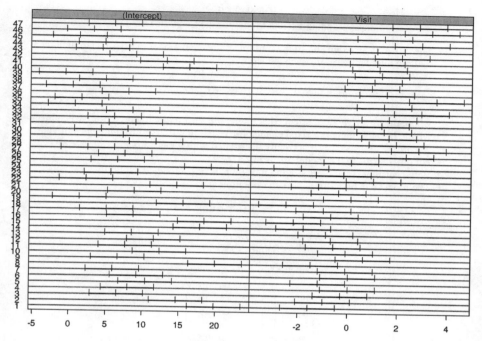

FIG. 7.8. Confidence intervals for intercepts and slopes of the linear regression of cognitive score on visit for each individual patient in the lecithin trial.

Eq. 7.2, ignores the repeated measures aspect of the data and incorrectly combines the between group and the within group variations in the residual standard error.

The model in Display 7.1, implying as it does the compound symmetry of the correlations between the cognitive scores on the five visits, is unlikely to capture the true correlation pattern suggested by Figure 7.7. One obvious extension of the model is to allow for a possible random effect of slope as well as of intercept, and such a model is described in Display 7.2. Note that this model allows for a more complex correlational pattern between the repeated measurements. The results of fitting this new model are given in Table 7.9.

Random effects models can be compared in a variety of ways, one of which is described in Display 7.3. For the two random effects models considered above, the AIC (Akaike information criterion) values are as follows.

Random intercepts only, AIC = 1282;
Random intercepts and random slopes, AIC = 1197.

Display 7.1
Simple Random Effects Model for the Lecithin Trial Data

- The model is an extension of the multiple regression model in Eq. (7.2) to include a random intercept term for each subject:

$$y_{ijk} = (\beta_0 + a_k) + \beta_1 \text{Visit}_j + \beta_2 \text{Group}_i + \epsilon_{ijk},$$

where the a_k are random effects that model the shift in the intercept for each subject, which, because there is a fixed change for visit, are preserved for all values of visit.
- The a_k are assumed to be normally distributed with mean zero and variance σ_a^2.
- All other terms in the model are as in Eq. (7.2).
- Between subject variability in the intercepts is now modeled explicitly.
- The model can be written in matrix notation as

$$\mathbf{y}_{ik} = \mathbf{X}_i \beta + \mathbf{Z} \mathbf{a}_k + \epsilon_{ik},$$

where

$$\beta' = [\beta_0, \beta_1, \beta_2],$$
$$\mathbf{y}'_{ik} = [y_{i1k}, y_{i2k}, y_{i3k}, y_{i4k}, y_{i5k}],$$
$$\mathbf{X}_i = \begin{pmatrix} 1 & 1 & \text{Group}_i \\ 1 & 2 & \text{Group}_i \\ 1 & 3 & \text{Group}_i \\ 1 & 4 & \text{Group}_i \\ 1 & 5 & \text{Group}_i \end{pmatrix},$$
$$\mathbf{Z}' = [1, 1, 1, 1, 1],$$
$$\epsilon'_{ik} = [\epsilon_{i1k}, \epsilon_{i2k}, \epsilon_{i3k}, \epsilon_{i4k}, \epsilon_{15k}].$$

- The presence of the random effect a_k that is common to each visit's observations for subject k allows the repeated measurements to have a covariance matrix of the following form:

$$\Sigma = \mathbf{Z} \sigma_a^2 \mathbf{Z}' + \sigma^2 \mathbf{I},$$

where $\mathbf{I}$ is a 5×5 identity matrix.
- This reduces to

$$\Sigma = \begin{pmatrix} \sigma^2 + \sigma_a^2 & \sigma_a^2 & \sigma_a^2 & \sigma_a^2 & \sigma_a^2 \\ \sigma_a^2 & \sigma^2 + \sigma_a^2 & \sigma_a^2 & \sigma_a^2 & \sigma_a^2 \\ \sigma_a^2 & \sigma_a^2 & \sigma^2 + \sigma_a^2 & \sigma_a^2 & \sigma_a^2 \\ \sigma_a^2 & \sigma_a^2 & \sigma_a^2 & \sigma^2 + \sigma_a^2 & \sigma_a^2 \\ \sigma_a^2 & \sigma_a^2 & \sigma_a^2 & \sigma_a^2 & \sigma^2 + \sigma_a^2 \end{pmatrix}.$$

- This is simply the compound symmetry pattern described in Chapter 5.
- The model can be fitted by maximum likelihood methods, as described in Brown and Prescott (1999).

TABLE 7.8

Results of Fitting the Model Described in Display 7.1 to the Data in Table 7.6

Parameter	Estimate	SE	t	p
β_0(Intercept)	6.927	0.916	7.56	<.001
β_1(Visit)	0.494	0.133	3.70	<.001
β_2(Group)	3.055	1.205	2.54	.015

Note. Random effects: $\hat{\sigma}_a = 3.89$, $\hat{\sigma} = 2.87$.

Display 7.2

Random Intercept, Random Slope Model for the Lecithin Trial Data

- The model in this case is

$$y_{ijk} = (\beta_0 + a_k) + (\beta_1 + b_k)\text{Visit}_j + \beta_2\text{Group}_i + \epsilon_{ijk}.$$

- A further random effect has been added compared with the model outlined in Display 7.1. The term b_k shifts the slope for each subject.
- The b_k are assumed to be normally distributed with mean zero and variance σ_b^2.
- The random effects are not necessarily assumed to be independent. A covariance parameter, σ_{ab}, is allowed in the model.
- All other terms in the model are as in Display 7.1.
- Between subject variability in both the intercepts and slopes is now modeled explicitly.
- Again, the model can be written in matrix notation as

$$\mathbf{y}_{ik} = \mathbf{X}_i\boldsymbol{\beta} + \mathbf{Z}\mathbf{b}_k + \boldsymbol{\epsilon}_{ik},$$

where now

$$\mathbf{Z} = \begin{pmatrix} 1 & 1 \\ 1 & 2 \\ 1 & 3 \\ 1 & 4 \\ 1 & 5 \end{pmatrix},$$

$$\mathbf{b}_k' = [a_k, b_k].$$

- The random effects are assumed to be from a bivariate normal distribution with both means zero and covariance matrix, $\boldsymbol{\Psi}$, given by

$$\boldsymbol{\Psi} = \begin{pmatrix} \sigma_a^2 & \sigma_{ab} \\ \sigma_{ab} & \sigma_b^2 \end{pmatrix}.$$

- The model implies the following covariance matrix for the repeated measures:

$$\boldsymbol{\Sigma} = \mathbf{Z}\boldsymbol{\Psi}\mathbf{Z}' + \sigma^2\mathbf{I}.$$

TABLE 7.9

Results of Fitting the Model Described in Display 7.2 to the Data in Table 7.6

Parameter	Estimate	SE	t	p
β_0(Intercept)	6.591	1.107	5.954	<.001
β_1(Visit)	0.494	0.226	2.185	.030
β_2(Group)	3.774	1.203	3.137	.003

Note. Random effects: $\hat{\sigma}_a = 6.22$, $\hat{\sigma}_b = 1.43$, $\hat{\sigma}_{ab} = -0.77$, $\hat{\sigma} = 1.76$.

Display 7.3

Comparing Random Effects Models for Longitudinal Data

- Comparisons of random effects models for longitudinal data usually involve as their basis values of the *log-likelihoods* of each model.
- The log-likelihood associated with a model arises from the parameter estimation procedure; see Brown and Prescott (1999) for details.
- Various indices are computed that try to balance goodness of fit with the number of parameters required to achieve the fit; the search for parsimony discussed in Chapter 1 is always remembered.
- One index is the *Akaike Information Criterion,* defined as

$$AIC = -2L + 2n_{par},$$

where L is the log-likelihood and n_{par} is the number of parameters in the model.
- Out of the models being compared, the one with the lowest AIC value is to be preferred.

It appears that the model in Display 7.2 is to be preferred for these data. This conclusion is reinforced by examination of Figures 7.9 and 7.10, the first of which shows the fitted regressions for each individual by using the random intercept model, and the second of which shows the corresponding predictions from the random intercepts and slopes model. Clearly, the first model does not capture the varying slopes of the regressions of an individual's cognitive score on the covariate visit. The number of words recalled in the lecithin treated group is about 4 more than in the placebo group, the 95% confidence interval is (1.37, 6.18).

As with the multiple regression model considered in Chapter 6, the fitting of random effects models to longitudinal data has to be followed by an examination of residuals to check for violations of assumptions. The situation is more complex now, however, because of the nature of the data—several residuals are available for each subject. Thus here the plot of standardized residuals versus fitted values from the random intercepts and slopes model given in Figure 7.11 includes a panel for each subject in the study. The plot does not appear to contain any

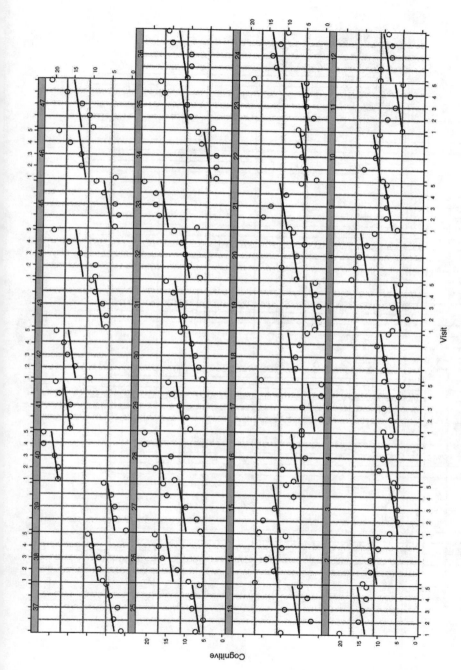

FIG. 7.9. Fitted regressions from the random intercepts model for each patient in the lecithin trial.

229

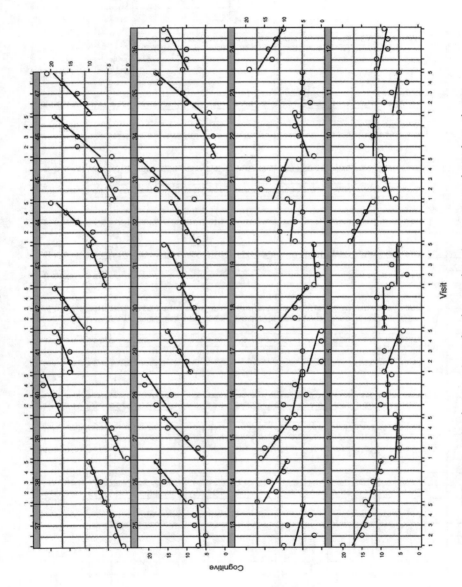

FIG. 7.10. Fitted regressions for the random intercepts and random slopes model for each patient in the lecithin trial.

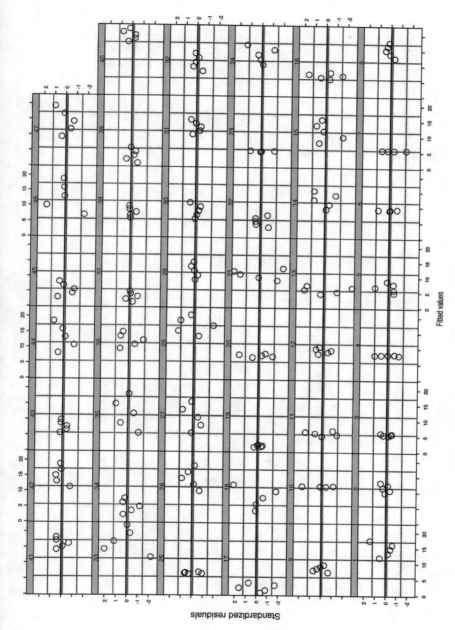

FIG. 7.11. Residuals from the random intercepts and random slopes model fitted to the lecithin trial data.

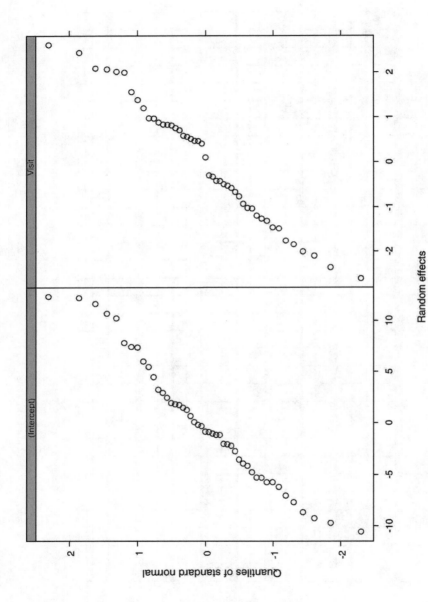

FIG. 7.12. Normal probability plots of the estimated random effects for intercepts and for slopes when the model in Display 7.2 is fitted to the lecithin trial data.

232

particularly disturbing features that might give cause for concern with the fitted model.

It may also be worth looking at the distributional properties of the estimated random effects from the model (such estimates are available from suitable random effects software such as that available in S-PLUS). Normal probability plots of the random intercept and random slope terms for the second model fitted to the lecithin data are shown in Figure 7.12. Again, there seem to be no particularly worrying departures from linearity in these plots.

7.4. SUMMARY

1. Longitudinal data are common in many areas, including the behavioral sciences, and they require particular care in their analysis.
2. The response feature, summary measure approach provides a simple, but not simplistic, procedure for analysis that can accommodate irregularly spaced observations and missing values.
3. Pretreatment values, if available, can be incorporated into a summary measures analysis in a number of ways, of which being used as a covariate in an ANCOVA is preferable.
4. The summary measure approach to the analysis of longitudinal data cannot give answers to questions about the development of the response over time or to whether this development differs in the levels of any between subject factors.
5. A more detailed analysis of longitudinal data requires the use of suitable models such as the random effects models briefly described in the chapter. Such models have been extensively developed over the past 5 years or so and are now routinely available in a number of statistical software packages.
6. In this chapter the models have been considered only for continuous, normally distributed responses, but they can be extended to deal with other types of response variable, for example, those that are categorical (see Brown and Prescott, 1999, for details).

COMPUTER HINTS

S-PLUS

In S-PLUS random effects models are available through the Statistics menu, by means of the following route.

1. Click **Statistics**, click **Mixed effects**, and click **Linear**, and the Mixed Effects dialog box appears.
2. Specify data set and details of the model that is to be fitted.

Random effect models can also be fitted by using the *lme* function in the command language approach, and in many respects this is to be preferred because many useful graphic procedures can then also be used. For example, if the lecithin trial data (see Table 7.6) are stored as a data frame, *lecit*, each row of which gives Subject number, Group label, Visit number, and Cognitive score, then the two random effects models considered in the text can be fitted and the results saved for further analysis by using the following commands.

$$lecit.f1 < -lme(Cognitive \sim Visit + Group, method$$

$$= ``ML", random = \sim 1|Subject, data = lecit)$$

$$lecit.f2 < -lme(Cognitive \sim Visit + Group, method$$

$$= ``ML", random = \sim Visit|Subject, data = lecit)$$

Many graphics are available for examining fits, and so on. Several have been used in the text. For example, to obtain the plots of predicted values from the two models fitted (see Figures 7.9 and 7.10), we can use the following commands.

$$plot(augPred(lecit.fit1), aspect = ``xy", grid = T),$$

$$plot(augPred(lecit.fit2), aspect = ``xy", grid = T).$$

EXERCISES

7.1. Apply the summary measure approach to the lecithin data in Table 7.6, using as your summary measure the maximum number of words remembered on any occasion.

7.2. Fit a random effects model with a random intercept and a random slope, and suitable fixed effects to the rat data in Table 7.2. Take care over coding the group to which a rat belongs.

7.3. The data in Table 7.10 arise from a trial of estrogen patches in the treatment of postnatal depression. Women who had suffered an episode of postnatal depression were randomly allocated to two groups; the members of one group received an estrogen patch, and the members of the other group received a dummy patch—the placebo. The dependent variable was a composite measure of depression, which was recorded on two occasions prior to randomization and for each of 4 months posttreatment. A number of observations are missing (indicated by -9).

TABLE 7.10
Estrogen Patch Trial Data

Group	Baseline 1	Baseline 2	Visit 1	Visit 2	Visit 3	Visit 4
0	18	18	17	18	15	17
0	25	27	26	23	18	17
0	19	16	17	14	−9	−9
0	24	17	14	23	17	13
0	19	15	12	10	8	4
0	22	20	19	11	9	8
0	28	16	13	13	9	7
0	24	28	26	27	−9	−9
0	27	28	26	24	19	14
0	18	25	9	12	15	12
1	21	21	13	12	9	9
1	27	27	8	17	15	7
1	24	15	8	12	10	10
1	28	24	14	14	13	12
1	19	15	15	16	11	14
1	17	17	9	5	3	6
1	21	20	7	7	7	12
1	18	18	8	1	1	2
1	24	28	11	7	3	2
1	21	21	7	8	6	6

Note. 0, placebo; 1, active groups. A −9 indicates a missing value.

1. Using the mean as a summary, compare the posttreatment measurements of the two groups by using the response feature procedure.

2. Calculate the change score defined as the difference in the mean of posttreatment values and the mean of pretreatment values, and test for a difference in average change score in the two groups.

3. Using the mean of the pretreatment values as a covariate and the mean of posttreatment values as the response, carry out an ANCOVA to assess posttreatment difference in the two groups.

4. Which of 2, or 3 would you recommend, and why?

7.4. Find a suitable random effects model for the data in Table 7.10 and interpret the results from the chosen model.

7.5. Fit a model to the lecithin data that includes random effects for both the intercept and slope of the regression of cognitive score on Visit and also includes fixed effects for Group and Group × Visit interaction.

7.6. For the random intercept and random slope model for the lecithin data described in Display 7.2, find the terms in the implied covariance matrix of the repeated measures, implicitly in terms of the parameters σ, σ_a, σ_b, and σ_{ab}.

8

Distribution-Free and Computationally Intensive Methods

8.1. INTRODUCTION

The statistical procedures described in the previous chapters have relied in the main for their validity on an underlying assumption of distributional normality. How well these procedures operate outside the confines of this normality constraint varies from setting to setting. Many people believe that in most situations, normal based methods are sufficiently powerful to make consideration of alternatives unnecessary. They may have a point; t tests and F tests, for example, have been shown to be relatively robust to departures from the normality assumption. Nevertheless, alternatives to normal based methods have been proposed and it is important for psychologists to be aware of these, because many are widely quoted in the psychological literature. One class of tests that do not rely on the normality assumptions are usually referred to as *nonparametric* or *distribution free*. (The two labels are almost synonymous, and we shall use the latter here.) Distribution-free procedures are generally based on the *ranks* of the raw data and are usually valid over a large class of distributions for the data, although they do often assume distributional symmetry. Although slightly less efficient than their normal theory competitors when the underlying populations *are* normal, they are often considerably more efficient when the underlying populations are *not* normal. A number of commonly used distribution-free procedures will be described in Sections 8.2–8.5.

A further class of procedures that do not require the normality assumption are those often referred to as *computationally intensive* for reasons that will become apparent in Sections 8.6 and 8.7. The methods to be described in these two sections use repeated permutations of the data, or repeated sampling from the data to generate an appropriate distribution for a test statistic under some null hypothesis of interest.

Although most of the work on distribution-free methods has concentrated on developing hypothesis testing facilities, it is also possible to construct confidence intervals for particular quantities of interest, as is demonstrated at particular points throughout this chapter.

8.2. THE WILCOXON–MANN–WHITNEY TEST AND WILCOXON'S SIGNED RANKS TEST

First, let us deal with the question of what's in a name. The statistical literature refers to two equivalent tests formulated in different ways as the *Wilcoxon rank sum test* and the *Mann–Whitney test*. The two names arise because of the independent development of the two equivalent tests by Wilcoxon (1945) and by Mann and Whitney (1947). In both cases, the authors' aim was to come up with a distribution-free alternative to the independent samples *t* test.

The main points to remember about the Wilcoxon–Mann–Whitney test are as follows.

1. The null hypothesis to be tested is that the two populations being compared have identical distributions. (For two normally distributed populations with common variance, this would be equivalent to the hypothesis that the means of the two populations are the same.)
2. The alternative hypothesis is that the population distributions differ in location (i.e., mean or median).
3. Samples of observations are available from each of the two populations being compared.
4. The test is based on the joint ranking of the observations from the two samples.
5. The test statistics is the sum of the ranks associated with one sample (the lower of the two sums is generally used).

Further details of the Wilcoxon–Mann–Whitney test are given in Display 8.1.

Display 8.1
Wilcoxon–Mann–Whitney Test

- Interest lies in testing the null hypothesis that two populations have the same probability distribution but the common distribution is not specified.
- The alternative hypothesis is that the population distributions differ in location.
- We assume that a sample of n_1 observations, $x_1, x_2, \ldots, x_{n_1}$, are available from the first population and a sample of n_2 observations, $y_1, y_2, \ldots, y_{n_2}$, from the second population.
- The combined sample of $n_1 + n_2$ observations are ranked and the test statistic, S, is the sum of the ranks of the observations from one of the samples (generally the lower of the two rank sums is used).
- If both samples come from the same population, a mixture of low, medium, and high ranks is to be expected in each. If, however, the alternative hypothesis is true, one of the samples would be expected to have lower ranks than the other.
- For small samples, tables giving p values are available in Hollander and Wolfe (1999), although the distribution of the test statistic can also now be found directly by using the *permutational* approach described in Section 8.6.
- A large sample approximation is also available for the Wilcoxon–Mann–Whitney test, which is suitable when n_1 and n_2 are both greater than 15. Under the null hypothesis the statistic Z given by

$$Z = \frac{S - n_1(n_1 + n_2 + 1)/2}{[n_1 n_2 (n_1 + n_2 + 1)/12]^{1/2}}$$

has approximately a standard normal distribution, and a p value can be assigned accordingly. (We are assuming that n_1 is the sample size associated with the sample's giving rise to the test statistic S.)
- If there are ties among the observations, then the tied observations are given the average of the ranks that they would be assigned if not tied. In this case the large sample approximation requires amending; see Hollander and Wolfe (1999) for details.

As our first illustration of the application of the test, we shall use the data shown in Table 8.1, adapted from Howell (1992). These data give the number of recent stressful life events reported by a group of cardiac patients in a local hospital and a control group of orthopedic patients in the same hospital. The result of applying the Wilcoxon–Mann–Whitney test to these data is a rank sum statistic of 21 with an associated p value of 0.13. There is no evidence that the number of stressful life events suffered by the two types of patient have different distributions.

To further illustrate the use of the Wilcoxon–Mann–Whitney test, we shall apply it to the data shown in Table 8.2. These data arise from a study in which the relationship between child-rearing practices and customs related to illness in several nonliterate cultures was examined. On the basis of ethnographical reports, 39 societies were each given a rating for the degree of typical and socialization anxiety, a concept derived from psychoanalytic theory relating to the severity and rapidity of oral socialization in child rearing. For each of the societies, a judgment

TABLE 8.1

Number of Stressful Life Events Among Cardiac
and Orthopedic Patients

Cardiac Patients (C)	Orthopedic Patients (O)
32 8 7 29 5 0	1 2 4 3 6

Observation	Rank
0	1
1	2 (O)
2	3 (O)
3	4 (O)
4	5 (O)
5	6
6	7 (O)
7	8
8	9
29	10
32	11

Note. Sum of "O ranks" is 21.

TABLE 8.2

Oral Socialization and Explanations of Illness: Oral
Socialization Anxiety Scores

Societies in which Oral Explanations of Illness are Absent

6 7 7 7 7 7 8 8 9 10 10 10 10 12 12 13

Societies in which Oral Explanations of Illness are Present

6 8 8 10 10 10 11 11 12 12 12 12 13 13 14 14 14 15 15 16 17

was also made (by an independent set of judges) of whether oral explanations of illness were present. Interest centers on assessing whether oral explanations of illness are more likely to be present in societies with high levels of oral socialization anxiety. This translates into whether oral socialization anxiety scores differ in location in the two types of societies. The result of the normal approximation test described in Display 8.1 is a Z value of -3.35 and an associated p value of .0008.

Display 8.2

Estimator and Confidence Interval Associated with the Wilcoxon–Mann–Whitney Test

- The estimator of the location difference, Δ, of the two populations is the median of the $n_1 \times n_2$ differences, $y_j - x_i$, of each pair of observations, made up of one observation from the first sample and one from the second sample.
- Constructing a confidence interval for Δ involves finding the two appropriate values among the ordered $n_1 \times n_2$ differences, $y_j - x_i$.
- For a symmetric two-sided confidence interval for Δ, with confidence level $1 - \alpha$, determine the upper $\alpha/2$ percentile point, $s_{\alpha/2}$, of the null distribution of S, either from the appropriate table in Hollander and Wolfe (1999), Table A.6, or from a permutational approach.
- Calculate

$$C_\alpha = \frac{n_1(2n_2 + n_1 + 1)}{2} + 1 - s_{\alpha/2}.$$

- The $1 - \alpha$ confidence interval (Δ_L, Δ_U) is then found from the values in the C_α (Δ_L) and the $n_1 n_2 + 1 - C_\alpha$ (Δ_U) positions in the ordered sample differences.

There is very strong evidence of a difference in oral socialization anxiety in the two types of societies.

The Wilcoxon–Mann–Whitney test is very likely to have been covered in the introductory statistics course enjoyed (suffered?) by many readers, generally in the context of testing a particular hypothesis, as described and illustrated above. It is unlikely, however, if such a course gave any time to consideration of associated estimation and confidence interval construction possibilities, although both remain as relevant to this distribution-free approach, as they are in applications of t tests and similar parametric procedures. Details of an estimator associated with the rank sum test and the construction of a relevant confidence interval are given in Display 8.2, and the results of applying both to the oral socialization anxiety data appear in Table 8.3. The estimated difference in location for the two types of societies is 3 units, with a 95% confidence interval of 2–6.

When samples from the two populations are matched or paired in some way, the appropriate distribution-free test becomes *Wilcoxon's signed-ranks test*. Details of the test are given in Display 8.3. As an example of the use of the test, it is applied to the data shown in Table 8.4. These data arise from an investigation of two types of electrodes used in some psychological experiments. The researcher was interested in assessing whether the two types of electrode performed similarly. Generally, a paired t test might be considered suitable for this situation, but here a normal probability plot of the differences of the observations on the two electrode types (see Figure 8.1) shows a distinct outlier, and so the signed-ranks test is to be preferred. The results of the large sample approximation test is a Z value of -2.20 with a p value of .03. There is some evidence of a difference between electrode

TABLE 8.3

Estimate of Location Difference and Confidence Interval Construction for Oral
Socialization Data

Differences Between Pairs of Observations, One Observation from Each Sample

---

```
0 2 2 4 4 5 5 6 6 6 6 7 7 7 8 8
8 9 9 10 11 −1 1 1 3 3 4 4 5 5 5 5
6 6 6 7 7 7 8 8 9 10 −1 1 1 3 3 4
4 5 5 5 5 6 6 6 7 7 7 8 8 9 10 −1
1 1 3 3 4 4 5 5 5 5 6 6 6 7 7 7
8 8 9 10 −1 1 1 3 3 4 4 5 5 5 5 6
6 6 7 7 7 8 8 9 10 −1 1 1 3 3 4 4
5 5 5 5 6 6 6 7 7 7 8 8 9 10 −2 0
0 2 2 3 3 4 4 4 4 5 5 5 6 6 6 7
7 8 9 −2 0 0 2 2 3 3 4 4 4 4 5 5
5 6 6 6 7 7 8 9 −3 −1 −1 1 1 2 2 3
3 3 3 4 4 4 5 5 5 6 6 7 8 −4 −2 −2
0 0 1 1 2 2 2 2 3 3 3 4 4 4 5 5
6 7 −4 −2 −2 0 0 1 1 2 2 2 2 3 3 3
4 4 4 5 5 6 7 −4 −2 −2 0 0 1 1 2 2
2 2 3 3 3 4 4 4 5 5 6 7 −4 −2 −2 0
0 1 1 2 2 2 2 3 3 3 4 4 4 5 5 6
7 −6 −4 −4 −2 −2 −1 −1 0 0 0 0 1 1 1 2
2 2 3 3 4 5 −6 −4 −4 −2 −2 −1 −1 0 0 0
0 1 1 1 2 2 2 3 3 4 5 −7 −5 −5 −3 −3
−2 −2 −1 −1 −1 −1 0 0 0 1 1 1 2 2 3 4
```

Note. The median of these differences is 3. The positions in the ordered
differences of the lower and upper limits of the 95% confidence interval are 117
and 252, leading to a confidence interval of (2,6).

types. Here, mainly as a result of the presence of the outlier, a paired t test gives a
p value of .09, implying no difference between the two electrodes. (Incidentally, the
outlier identified in Figure 8.1 was put down by the experimenter to the excessively
hairy arms of one subject!)

Again you may have met the signed-ranks test on your introductory statistics course, but perhaps not the associated estimator and confidence interval,
both of which are described in Display 8.4. For the electrode, data application
of these procedures gives an estimate of −65 and an approximate 95% confidence interval of −537.0 to −82.5. Details of the calculations are shown in
Table 8.5.

Display 8.3
Wilcoxon's Signed-Ranks Test

- Assume we have two observations, x_i and y_i, on each of n subjects in our sample, for example, before and after treatment. We first calculate the differences $z_i = x_i - y_i$ between each pair of observations.
- To compute the Wilcoxon signed-rank statistic, T^+, form the absolute values of the differences, z_i, and then order them from least to greatest.
- Now assign a positive or negative sign to the ranks of the differences according to whether the corresponding difference was positive or negative. (Zero values are discarded, and the sample size n altered accordingly.)
- The statistic T^+ is the sum of the positive signed ranks. Tables are available for assigning p values; see Table A.4 in Hollander and Wolfe (1999).
- A large sample approximation involves testing the statistic Z as a standard normal:

$$Z = \frac{T^+ - n(n+1)/4}{[n(n+1)(2n+1)/24]^{1/2}}.$$

- If there are ties among the calculated differences, assign each of the observations in a tied group the average of the integer ranks that are associated with the tied group.

TABLE 8.4
Skin Resistance and Electrode Type

Subject	Electrode 1	Electrode 2
1	500	400
2	660	600
3	250	370
4	72	140
5	135	300
6	27	84
7	100	50
8	105	180
9	90	180
10	200	290
11	15	45
12	160	200
13	250	400
14	170	310
15	66	1000
16	107	48

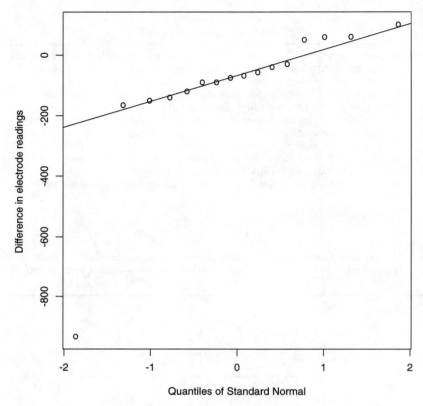

FIG. 8.1. Probability plot for differences between electrode readings for the data in Table 8.4.

Display 8.4
Estimator and Confidence Interval Associated with Wilcoxon's Signed-Rank Statistic

- An estimator of the treatment effect, θ, is the median of the $n(n+1)/2$ averages of pairs of the n differences on which the statistic T^+ is based, that is, the averages $(z_i + z_j)/2$ for $i \leq j = 1, \ldots, n$.
- For a symmetric two-sided confidence interval for θ, with confidence level $1 - \alpha$, we first find the upper $\alpha/2$ percentile point of the null distribution of T^+, $t_{\alpha/2}$, from Table A.4 in Hollander and Wolfe (1999) or from a permutational computation. Then set

$$C_\alpha = \frac{n(n+1)}{2} + 1 - t_{\alpha/2}.$$

- The lower and upper limits of the required confidence interval are now found as the C_α and $t_{\alpha/2}$ position values of the ordered averages of pairs of differences.

TABLE 8.5
Estimation and Confidence Interval Construction for Electrode Data

---

Averages of Pairs of Sample Differences

100.0 80.0 −10.0 16.0 −32.5 21.5 75.0 12.5 5.0
5.0 35.0 30.0 −25.0 −20.0 −417.0 79.5 80.0 60.0
−30.0 −4.0 −52.5 1.5 55.0 −7.5 −15.0 −15.0 15.0
10.0 −45.0 −40.0 −437.0 59.5 −10.0 −30.0 −120.0 −94.0
−142.5 −88.5 −35.0 −97.5 −105.0 −105.0 −75.0 −80.0 −135.0
−130.0 −527.0 −30.5 16.0 −4.0 −94.0 −68.0 −116.5 −62.5
−9.0 −71.5 −79.0 −79.0 −49.0 −54.0 −109.0 −104.0 −501.0
−4.5 −32.5 −52.5 −142.5 −116.5 −165.0 −111.0 −57.5 −120.0
−127.5 −127.5 −97.5 −102.5 −157.5 −152.5 −549.5 −53.0 21.5
1.5 −88.5 −62.5 −111.0 −57.0 −3.5 −66.0 −73.5 −73.5
−43.5 −48.5 −103.5 −98.5 −495.5 1.0 75.0 55.0 −35.0
−9.0 −57.5 −3.5 50.0 −12.5 −20.0 −20.0 10.0 5.0
−50.0 −45.0 −442.0 54.5 12.5 −7.5 −97.5 −71.5 −120.0
−66.0 −12.5 −75.0 −82.5 −82.5 −52.5 −57.5 −112.5 −107.5
−504.5 −8.0 5.0 −15.0 −105.0 −79.0 −127.5 −73.5 −20.0
−82.5 −90.0 −90.0 −60.0 −65.0 −120.0 −115.0 −512.0 −15.5
5.0 −15.0 −105.0 −79.0 −127.5 −73.5 −20.0 −82.5 −90.0
−90.0 −60.0 −65.0 −120.0 −115.0 −512.0 −15.5 35.0 15.0
−75.0 −49.0 −97.5 −43.5 10.0 −52.5 −60.0 −60.0 −30.0
−35.0 −90.0 −85.0 −482.0 14.5 30.0 10.0 −80.0 −54.0
−102.5 −48.5 5.0 −57.5 −65.0 −65.0 −35.0 −40.0 −95.0
−90.0 −487.0 9.5 −25.0 −45.0 −135.0 −109.0 −157.5 −103.5
−50.0 −112.5 −120.0 −120.0 −90.0 −95.0 −150.0 −145.0 −542.0
−45.5 −20.0 −40.0 −130.0 −104.0 −152.5 −98.5 −45.0 −107.5
−115.0 −115.0 −85.0 −90.0 −145.0 −140.0 −537.0 −40.5 −417.0
−437.0 −527.0 −501.0 −549.5 −495.5 −442.0 −504.5 −512.0 −512.0
−482.0 −487.0 −542.0 −537.0 −934.0 −437.5 79.5 59.5 −30.5
−4.5 −53.0 1.0 54.5 −8.0 −15.5 −15.5 14.5 9.5
−45.5 −40.5 −437.5 59.0

Note. The median of these averages is −65. The upper and lower limits
of the 95% confidence interval are in position 7 and 107 of the ordered
values of the averages, leading to the interval (−537.0, −82.5).

8.3. DISTRIBUTION-FREE TEST
FOR A ONE-WAY DESIGN
WITH MORE THAN TWO GROUPS

Just as the Wilcoxon–Mann–Whitney test and the Wilcoxon signed-ranks test can
be considered as distribution-free analogs of the independent samples and paired
samples *t* tests, the *Kruskal–Wallis procedure* is the distribution-free equivalent of

Display 8.5

Kruskal–Wallis Distribution-Free Procedure for One-Way Designs

---

- Assume there are k populations to be compared and that a sample of n_j observations is available from population j, $j = 1, \ldots, k$.
- The hypothesis to be tested is that all the populations have the same probability distribution.
- For the Kruskal–Wallis test to be performed, the observations are first ranked without regard to group membership and then the sums of the ranks of the observations in each group are calculated. These sums will be denoted by $R_1, R_2, \ldots, R_k$.
- If the null hypothesis is true, we would expect the R_js to be more or less equal, apart from differences caused by the different sample sizes.
- A measure of the degree to which the R_js differ from one another is given by

$$H = \frac{12}{N(N+1)} \sum_{j=1}^{k} \frac{R_j^2}{n_j} - 3(N+1),$$

where $N = \sum_{j=1}^{k} n_j$
- Under the null hypothesis the statistic H has a chi-squared distribution with $k - 1$ degrees of freedom.

---

the one-way ANOVA procedure described in Chapter 3. Details of the Kruskal–Wallis method for one-way designs with three or more groups are given in Display 8.5. (When the number of groups is two, the Kruskal–Wallis test is equivalent to the Wilcoxon–Mann–Whitney test.)

To illustrate the use of the Kruskal–Wallis method, we apply it to the data shown in Table 8.6. These data arise from an investigation of the possible beneficial effects of the pretherapy training of clients on the process and outcome of counseling and psychotherapy. Sauber (1971) investigated four different approaches to pretherapy training.

- Control (no treatment),
- Therapeutic reading (TR) (indirect learning),
- Vicarious therapy pretraining (VTP) (videotaped, vicarious learning), and
- Group, role induction interview (RII) (direct learning).

Nine clients were assigned to each of these four conditions and a measure of psychotherapeutic attraction eventually given to each client. Applying the Kruskal–Wallis procedure to these data gives a chi-squared test statistic of 4.26, which with 3 degrees of freedom has an associated p value of .23. There is no evidence of a difference between the four pretherapy regimes.

(Hollander and Wolfe, 1999, describe some distribution-free analogs to the multiple comparison procedures discussed in Chapter 3, for identifying which

TABLE 8.6
Psychotherapeutic Attraction Scores for Four Experimental Conditions

Control	Reading (TR)	Videotape (VTP)	Group (RII)
0	0	0	1
1	6	5	5
3	7	8	12
3	9	9	13
5	11	11	19
10	13	13	22
13	20	16	25
17	20	17	27
26	24	20	29

Display 8.6
Friedman's Rank Test for Correlated Samples

- Here we assume that a sample of n subjects are observed under k conditions.
- First the k observations for each subject are ranked, to give the values r_{ij}, $i = 1, \ldots, n$, $j = 1, \ldots, k$.
- Then the sums of these ranks over conditions are found as $R_j = \sum_{i=1}^{n} r_{ij}$, $j = 1, \ldots, k$.
- The Friedman statistic S is then given by

$$S = \frac{12}{nk(k+1)} \sum_{j=1}^{k} R_j^2 - 3n(k+1).$$

- Under the null hypothesis that in the population the conditions have the same distribution, S has a chi-squared distribution with $k - 1$ degrees of freedom.

particular groups in the one-way design differ, after obtaining a significant result from the Kruskal–Wallis test.)

8.4. DISTRIBUTION-FREE TEST FOR A ONE-WAY REPEATED MEASURES DESIGN

When the one-way design involves repeated measurements on the same subject, a distribution-free test from Friedman (1937) can be used. The test is described in Display 8.6. The test is clearly related to a standard repeated measures ANOVA applied to ranks rather than raw scores.

TABLE 8.7
Percentage of Consonants Correctly Identified under Each of Seven Conditions

Subject	A	L	AL	C	AC	LC	ALC
1	1.1	36.9	52.4	42.9	31.0	83.3	63.0
2	1.1	33.3	34.5	34.5	41.7	77.3	81.0
3	13.0	28.6	40.5	33.3	44.0	81.0	76.1
4	0	23.8	22.6	33.3	33.3	69.0	65.5
5	11.9	40.5	57.1	35.7	46.4	98.8	96.4
6	0	27.4	46.4	42.9	47.4	78.6	77.4
7	5.0	20.2	22.6	35.7	37.0	69.0	73.8
8	4.0	29.8	42.9	13.0	33.3	95.2	91.7
9	0	27.4	38.0	42.9	45.2	89.3	85.7
10	1.1	26.2	31.0	31.0	32.1	70.2	71.4
11	2.4	29.8	38.0	34.5	46.4	86.9	92.9
12	0	21.4	21.4	41.7	33.3	67.9	59.5
13	0	32.1	33.3	44.0	34.5	86.9	82.1
14	0	28.6	23.8	32.1	39.3	85.7	72.6
15	1.1	28.6	29.8	41.7	35.7	81.0	78.6
16	1.1	36.9	33.3	25.0	31.0	95.2	95.2
17	0	27.4	26.1	40.5	44.0	91.7	89.3
18	0	41.7	35.7	42.9	45.2	95.2	95.2

To illustrate the use of the Friedman method, we apply it to the data shown in Table 8.7, taken from a study reported by Nicholls and Ling (1982), into the effectiveness of a system using hand cues in the teaching of language to severely hearing-impaired children. In particular, they considered syllables presented to hearing-impaired children under the following seven conditions.

A: audition,
L: lip reading,
AL: audition and lip reading,
C: cued speech,
AC: audition and cued speech,
LC: lip reading and cued speech, and
ALC: audition, lip reading, and cued speech.

The 18 subjects in the study were all severely hearing-impaired children who had been taught through the use of cued speech for at least 4 years. Syllables were presented to the subjects under each of the seven conditions (presented in random orders), and the subjects were asked in each case to identify the consonants in each syllable by writing down what they perceived them to be. The subjects' results were scored by marking properly identified consonants in the appropriate order as correct. Finally, an overall percentage correct was assigned to each participant under each experimental condition; it is these percentages that are shown in Table 8.7.

Applying the Friedman procedure to the data in Table 8.7 gives the chi-squared statistic of 94.36 with 6 degrees of freedom. The associated p value is very small, and there is clearly a difference in the percentage of consonants identified in the different conditions.

(Possible distribution-free approaches to more complex repeated measures situations are described in Davis, 1991.)

8.5. DISTRIBUTION-FREE CORRELATION AND REGRESSION

A frequent requirement in many psychological studies is to measure the correlation between two variables observed on a sample of subjects. In Table 8.8, for example, examination scores (out of 75) and corresponding exam completion times (seconds) are given for 10 students. A scatterplot of the data is shown in Figure 8.2.

The usual Pearson's product moment correlation coefficient, r, takes the value 0.50, and a t test that the population correlation coefficient, ρ, takes the value zero gives $t = 1.63$. (Readers are reminded that the test statistic here is $r\sqrt{(n-2)/(1-r^2)}$, which under the null hypothesis that $\rho = 0$ has a t distribution with $n - 1$ degrees of freedom.) Dealing with Pearson's coefficient involves

TABLE 8.8
Examination Scores and Time to Complete
Examination

Score	Time (s)
49	2160
70	2063
55	2013
52	2000
61	1420
65	1934
57	1519
71	2735
69	2329
44	1590

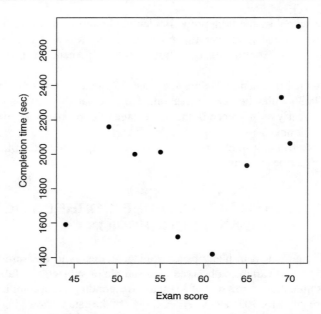

FIG. 8.2. Scatterplot of examination scores and completion times.

the assumption that the data arises from a bivariate normal distribution (see Everitt and Wykes, 1999, for a definition). But how can we test for independence when we are not willing to make this assumption? There are a number of possibilities.

8.5.1. Kendall's Tau

Kendall's tau is a measure of correlation that can be used to assess the independence of two variables, without assuming that the underlying bivariate distribution is bivariate normal. Details of the calculation and testing of the coefficient are given in Display 8.7. For the data in Table 8.8, Kendall's tau takes the value 0.29, and the large sample z test is 1.16 with an associated p value of 0.24. The estimated correlation is lower than the value given by using Pearson's coefficient, although the associated tests for independence both conclude that examination score and completion time are independent.

8.5.2. Spearman's Rank Correlation

Another commonly used distribution-free correlation measure is Spearman's coefficient. Details of its calculation and testing are given in Display 8.8. For the examination scores and times data, it takes the value 0.42, and the large sample

Display 8.7

Kendall's Tau

- Assume n observations on two variables are available. Within each variable the observations are ranked.
- The value of Kendall's tau is based on the number of *inversions* in the two sets of ranks. This term can best be explained with the help of a small example involving three subjects with scores on two variables, which after ranking give the following;

Subject	Variable 1	Variable 2
1	1	1
2	2	3
3	3	2

When the subjects are listed in the order of their ranks on variable 1, there is an inversion of the ranks on variable 2 (rank 3 appears before rank 2).
- Kendall's tau statistic can now be defined as

$$\tau = 1 - \frac{2I}{n(n-1)/2},$$

where I is the number of inversions in the data.
- A significance test for the null hypothesis that the population correlation is zero is provided by z given by

$$z = \frac{\tau}{\sqrt{[2(2n+5)]/[9n(n-1)]}},$$

which can be tested as a standard normal.

Display 8.8

Spearman's Correlation Coefficient for Ranked Data

- Assume that n observations on two variables are available. Within each variable the observations are ranked.
- Spearman's correlation coefficient is defined as

$$r_S = 1 - \frac{6\sum_{j=1}^{n} D_j^2}{n(n^2-1)},$$

where D_j is the difference in the ranks assigned to subject j on the two variables. (This formula arises simply from applying the usual Pearson correlation coefficient formula to the ranked data.)
- A significance test for the null hypothesis that the population correlation is zero is given by

$$z = (n-1)^{1/2} r_S,$$

which is tested as a standard normal.

---

z test is 1.24 with an associated p value of 0.22. Again, the conclusion is that the two variables are independent.

8.5.3. Distribution-Free Simple Linear Regression

Linear regression, as described in Chapter 6, is one of the most commonly used of statistical procedures. Estimation and testing of regression coefficients depends largely on the assumption that the error terms in the model are normally distributed. It is, however, possible to use a distribution-free approach to simple regression where there is a single explanatory variable, based on a method suggested by Theil (1950). Details are given in Display 8.9.

We shall use the distribution-free regression procedure on the data introduced in Chapter 6 (see Table 6.1), giving the average number of words known by children

Display 8.9
Distribution-Free Simple Linear Regression

- The null hypothesis of interest is that the slope of the regression line between two variables x and y is zero.
- We have a sample of n observation on the two variables, $x_j, y_j, \ j = 1, \ldots, n$.
- The test statistic, C, is given by

$$C = \sum_{i=1}^{n} \sum_{j=i+1}^{n} c(y_j - y_i),$$

where the function c is zero if $y_j = y_i$, takes the value 1 if $y_j > y_i$, and the value -1 if $y_j < y_i$.
- The p value of the test statistic can be found from Table A.30 in Hollander and Wolfe (1999).
- A large sample approximation uses the statistic z given by

$$z = \frac{C}{[n(n - 1)(2n + 5)/18]^{1/2}},$$

referred to a standard normal distribution.
- An estimator of the regression slope is obtained as

$$\hat{\beta} = \text{median}\{S_{ij}, 1 \leq i < j \leq n\},$$

where $S_{ij} = (y_j - y_i)/(x_j - x_i)$.
- For a symmetric two-sided confidence interval for β with confidence level $1 - \alpha$, we first obtain the upper $\alpha/2$ percentile point, $k_{\alpha/2}$ of the null distribution of C from Table A.30 of Hollander and Wolfe (1999).
- The lower and upper limits of the required confidence interval are found at the Mth and Qth positions of the ordered sample values S_{ij}, where

$$M = \frac{n - k_{\alpha/2} - 2}{2},$$
$$Q = M + k_{\alpha/2} - 2.$$

TABLE 8.9

Distribution-Free Estimation for Regression Coefficient of Vocabulary
Scores on Age

S_{ij} *Values (see Display 8.9) for the Vocabulary Data*

38.00000	269.00000	295.33333	446.50000	487.60000	512.33333
533.42857	517.25000	511.80000	500.00000	424.00000	582.66667
600.00000	607.20000	616.00000	585.71429	564.44444	348.00000
624.00000	633.33333	634.00000	639.20000	600.00000	572.50000
900.00000	776.00000	729.33333	712.00000	650.40000	604.57143
652.00000	644.00000	649.33333	588.00000	555.33333	636.00000
648.00000	566.66667	536.00000	660.00000	532.00000	511.00000
404.00000	461.33333	490.00000			

Note. The estimates of the intercept and the slope are obtained from these
values as

$$\hat{\alpha} = \text{median}(\text{vocabulary score} - \hat{\beta} \times \text{age});$$
$$\hat{\beta} = \text{median}\{S_{ij}\}.$$

This leads to the values $\hat{\alpha} = -846.67$, $\hat{\beta} = 582.67$. The test statistic, C,
for testing the null hypothesis that the slope is zero takes the value 45. The
associated p values for the appropriate table in Hollander and Wolfe (1999)
is very small, and so we can (not surprisingly) reject the hypothesis that the
regression coefficient for age and vocabulary score is not zero.

at various ages. Details of the calculations and results are shown in Table 8.9. The
fitted distribution-free regression line and, for comparison, that obtained by using
least squares are shown on a scatterplot of the data in Figure 8.3. Here the two
fitted lines are extremely similar.

Other aspects of distribution-free regression including a procedure for multiple
regression are discussed in Hollander and Wolfe (1999). Such methods based on
what might be termed a *classical* distribution-free approach, that is, using ranks
in some way, are probably *not* of great practical importance. However, the more
recent developments such as *locally weighted regression* and *spline smoothers*,
which allow the data themselves to suggest the form of the regression relationship
between variables, are of increasing importance. Interested readers can consult
Cleveland (1985) for details.

8.6. PERMUTATION TESTS

The Wilcoxon–Mann–Whitney test and other distribution-free procedures described in the previous section are all simple examples of a general class of tests
known as either *permutation* or *randomization* tests. Such procedures were first

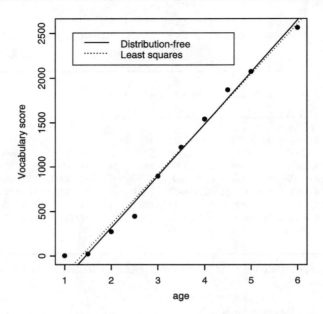

FIG. 8.3. Scatterplot of vocabulary scores at different ages, showing fitted least squares and distribution-free regressions.

introduced in the 1930s by Fisher and Pitman, but initially they were largely of theoretical rather than practical interest because of the lack of the computer technology required to undertake the extensive computation often needed in their application. However, with each increase in computer speed and power, the permutation approach is being applied to a wider and wider variety of problems, and with today's more powerful generation of personal computers, it is often faster to calculate a p value for an exact permutation test than to look up an asymptotic approximation in a book of tables. Additionally, the statistician (or the psychologist) is not limited by the availability of tables but is free to choose a test statistic exactly matched to testing a particular null hypothesis against a particular alternative. Significance levels are then, so to speak, "computed on the fly" (Good, 1994). The stages in a general permutation test are as follows.

1. Choose a test statistic S.
2. Compute S for the original set of observations.
3. Obtain the *permutation distribution* of S by repeatedly rearranging the observations. When two or more samples are involved (e.g., when the difference between groups is assessed), all the observations are combined into a single large sample before they are rearranged.
4. For a chosen significance level α, obtain the upper α-percentage point of the permutational distribution and accept or reject the null hypothesis

according to whether the value of S calculated for the original observations is smaller or larger than this value. The resultant p value is often referred to as *exact*.

To illustrate the application of a permutation test, consider a situation involving two treatments in which three observations of some dependent variable of interest are made under each treatment. Suppose that the observed values are as follows.

Treatment 1:   121, 118, 110,
Treatment 2:   34, 33, 12.

The null hypothesis is that there is no difference between the treatments in their effect on the dependent variable. The alternative is that the first treatment results in higher values of the dependent variable. (The author is aware that the result in this case is pretty clear without any test!)

The first step in a permutation test is to choose a test statistic that discriminates between the hypothesis and the alternative. An obvious candidate is the sum of the observations for the first treatment group. If the alternative hypothesis is true, this sum ought to be larger than the sum of the observations in the second treatment group. If the null hypothesis is true, then the sum of the observations in each group should be approximately the same. One sum might be smaller or larger than the other by chance, but the two should not be very different.

The value of the chosen test statistic for the observed values is $121 + 118 + 110 = 349$. To generate the necessary permutation distribution, remember that under the null hypothesis the labels "treatment 1" and "treatment 2" provide no information about the test statistic, as the observations are expected to have almost the same values in each of the two treatment groups. Consequently, to permutate the observations, simply reassign the six labels, three treatment 1 and three treatment 2, to the six observations; for example, treatment 1—121, 118, 34; treatment 2—110, 12, 22; and so on. Repeat the process until all the possible 20 distinct arrangements have been tabulated as shown in Table 8.10.

From the results given in Table 8.10, it is seen that the sum of the observations in the original treatment 1 group, that is, 349, is equalled only once and never exceeded in the distinct random relabelings. If chance alone is operating, then such an extreme value has a 1 in 20 chance of occurring, that is, 5%. Therefore at the conventional .05 significance level, the test leads to the rejection of the null hypothesis in favor of the alternative.

The Wilcoxon–Mann–Whitney test described in the previous section is also an example of a permutation test, but one applied to the ranks of the observations rather than their actual values. Originally the advantage of this procedure was that, because ranks always take the same values (1, 2, etc.), previously tabulated distributions could be used to derive p values, at least for small samples. Consequently, lengthy computations were avoided. However, it is now relatively simple

TABLE 8.10
Permutation Distribution of Example with Three Observations in Each Group

	First Group			Second Group			Sum of First Group Obs.
1	121	118	110	34	22	12	349
2	121	118	34	110	22	12	273
3	121	110	34	118	22	12	265
4	118	110	34	121	22	12	262
5	121	118	22	110	34	12	261
6	121	110	22	118	34	12	253
7	121	118	12	110	34	22	251
8	118	110	22	121	34	12	250
9	121	110	12	118	34	22	243
10	118	110	12	121	34	22	240
11	121	34	22	118	110	12	177
12	118	34	22	121	110	12	174
13	121	34	12	118	110	22	167
14	110	34	22	121	118	12	166
15	118	34	12	121	110	22	164
16	110	34	12	121	118	22	156
17	121	22	12	118	110	34	155
18	118	22	12	121	110	34	152
19	110	22	12	121	118	34	144
20	34	22	12	121	118	110	68

TABLE 8.11
Data from a Study of Organization and Memory

Training Group	No Training Group
0.35 0.40 0.41 0.46 0.49	0.33 0.34 0.35 0.37 0.39

to calculate the required permutation distribution, a point that will be demonstrated by using the data shown in Table 8.11, which results from a study of organization in the memory of mildly retarded children attending a special school. A bidirectional subjective organizational measure (SO2) was used to assess the amount of intertrial consistency in a memory task given to children in two groups, one of which had received a considerable amount of training in sorting tasks. The question of interest here is whether there is any evidence of an increase in organization in the group

TABLE 8.12
Permutational Distribution of the Wilcoxon–Mann–Whitney Rank Sum Statistic
Applied to the SO2 Scores

- There are a total of 252 possible permutations of pupils to groups, that is ($\binom{10}{5}$).
- The distribution of the sum of the ranks in the training group for all possible 252 permutations is as follows.

S	15	16	17	18	19	20	21	22	23	24	25	26	27
f	1	1	2	3	5	7	9	11	14	16	18	19	20
S	28	29	30	31	32	33	34	35	36	37	38	39	40
f	20	19	18	16	14	11	9	7	5	3	2	1	1

- So, for example, there is one arrangement where the ranks sum to 15, one where they sum to 16, two where they sum to 17, and so on.
- From this distribution we can determine the probability of finding a value of the sum of the ranks equal to or greater than the observed value of 37, if the null hypothesis is true:

$$\Pr(S \geq 37) = (3 + 2 + 1 + 1)/(252) = 0.028.$$

- It is this type of calculation that gives the values in the tables for assigning p values to values of the Wilcoxon–Mann–Whitney rank sum test.

that had received training (more details of the study are given in Robertson, 1991.) The permutational distribution of the Wilcoxon–Mann–Whitney rank sum statistic is shown in Table 8.12. As can be seen, this leads to an exact p value for the test of a group difference on the SO2 scores of .028. There *is* evidence of a training effect.

8.7. THE BOOTSTRAP

The *bootstrap* is a data-based method for statistical inference. Its introduction into statistics is relatively recent, because the method is computationally intensive. According to Efron and Tibshirani (1993), the term *bootstrap* derives from the phrase "to pull oneself up by one's bootstraps," widely considered to be based on the eighteenth century adventures of Baron Munchausen by Rudolph Erich Raspe. (The relevant adventure is the one in which the Baron had fallen into the bottom of a deep lake. Just when it looked as if all was lost, he thought he could pick himself up by his own bootstraps.)

The bootstrap looks like the permutational approach in many respects, requires a minimal number of assumptions for its applications, and derives critical values for testing and constructing confidence intervals from the data at hand. The stages in the bootstrap approach are as follows.

1. Choose a test statistic S.
2. Calculate S for the original set of observations.

3. Obtain the *bootstrap distribution* of S by repeatedly resampling from the observations. In the multigroup situation, samples are not combined but are resampled separately. Sampling is *with* replacement.
4. Obtain the upper α-percentage point of the bootstrap distribution and accept or reject the null hypothesis according to whether S for the original observations is smaller or larger than this value.
5. Alternatively, construct a confidence interval for the test statistic by using the bootstrap distribution (see the example given later).

Unlike a permutation test, the bootstrap does not provide exact p values. In addition, it is generally less powerful than a permutation test. However, it may be possible to use a bootstrap procedure when no other statistical method is applicable. Full details of the bootstrap, including many fascinating examples of its application, are given by Efron and Tibshirani (1993). Here, we merely illustrate its use in two particular examples, first in the construction of confidence intervals for the difference in means and in medians of the groups involved in the WISC data introduced in Chapter 3 (see Table 3.6), and then in regression, using the vocabulary scores data used earlier in this chapter and given in Chapter 6 (see Table 6.1).

The results for the WISC example are summarized in Table 8.13. The confidence interval for the mean difference obtained by using the bootstrap is seen to be narrower than the corresponding interval obtained conventionally with the t statistic.

TABLE 8.13
Construction of Confidence Intervals for the WISC Data by Using the Bootstrap

---

- The procedure is based on drawing random samples of 16 observations with replacement from each of the row and corner groups.
- The random samples are found by labeling the observations in each group with the integers $1, 2, \ldots, 16$ and selecting random samples of these integers (with replacement), using an appropriate computer algorithm (these are generally available in most statistical software packages).
- The mean difference and median difference is calculated for each bootstrap sample.
- In this study, 1000 bootstrap samples were used, resulting in 1000 mean and median differences.
- The first five bootstrap samples in terms of the integers $1, 2, \ldots, 16$ were as follows.

    1, 2, 2, 6, 7, 7, 8, 9, 11, 11, 13, 13, 14, 15, 16, 16
    1, 2, 3, 3, 5, 6, 7, 10, 11, 11, 12, 13, 14, 15, 16
    1, 1, 3, 3, 3, 4, 5, 6, 8, 8, 9, 14, 14, 15, 16, 16
    1, 2, 4, 4, 4, 5, 5, 5, 6, 10, 11, 11, 12, 12, 12, 14
    1, 2, 3, 3, 5, 5, 6, 6, 9, 9, 12, 13, 14, 15, 15, 16

- A rough 95% confidence interval can be derived by taking the 25th and 975th largest of the replicates, leading to the following intervals: mean $(16.06, 207.81)$, median $(7.5, 193.0)$.
- The 95% confidence interval for the mean difference derived in the usual way, assuming normality of the population distributions and homogeneity of variance, is $(11.65, 211.59)$.

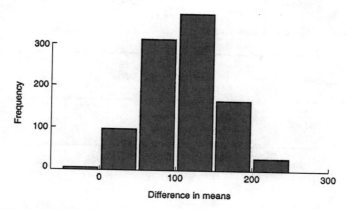

FIG. 8.4. Histogram of mean differences for 1000 bootstrap samples of the WISC data.

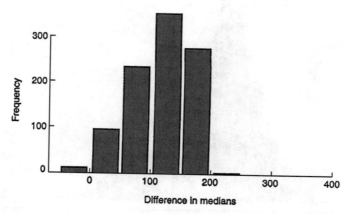

FIG. 8.5. Histogram of median differences for 1000 bootstrap samples of the WISC data.

Histograms of the mean and median differences obtained from the bootstrap samples are shown in Figures 8.4 and 8.5.

The bootstrap results for the regression of vocabulary scores on age are summarized in Table 8.14. The bootstrap confidence interval of (498.14, 617.03) is wider than the interval given in Chapter 6 based on the assumption of normality (513.35, 610.51). The bootstrap results are represented graphically in Figures 8.6 and 8.7. We see that the regression coefficients calculated from the bootstrap samples show a minor degree of skewness, and that the regression coefficient calculated from the observed data is perhaps a little biased compared to the mean of the bootstrap distribution of 568.35.

TABLE 8.14
Bootstrap Results for Regression of Vocabulary Scores on Age

Observed Value	Bias	Mean	SE
561.93	6.42	568.35	29.65

Number of bootstrap samples, 1000; 95% Confidence interval (498.14, 617.03).

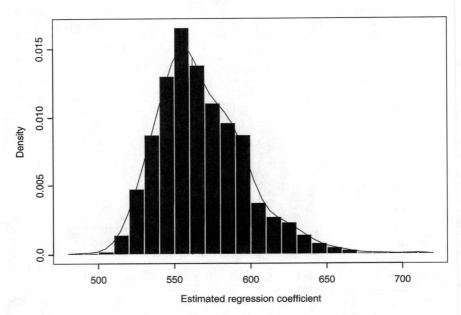

FIG. 8.6. Histogram of regression coefficients of vocabulary scores on age for 1000 bootstrap samples.

8.8. SUMMARY

1. Distribution-free tests are useful alternatives to parametric approaches, when the sample size is small and therefore evidence for any distributional assumption is not available empirically.
2. Such tests generally operate on ranks and so are invariant under transformations of the data that preserve order. They use only the ordinal property of the raw data.

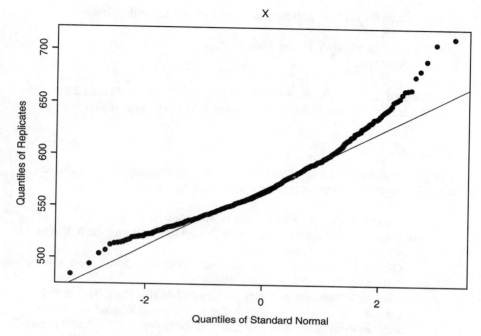

FIG. 8.7. Normal probability plot of bootstrap regression coefficients for vocabulary scores data.

3. Permutation tests and the bootstrap offer alternative approaches to making distribution-free inferences. Both methods are computationally intensive.
4. This chapter has given only a brief account of this increasingly important area of modern applied statistics. Comprehensive accounts are given by Good (1994) and by Efron and Tibshirani (1993).

COMPUTER HINTS

SPSS

Many distribution-free tests are available in SPSS; for example, to apply the Wilcoxon–Mann–Whitney test, we would use the following steps.

1. Click on **Statistics**, click on **Nonparametric Tests**, and then click on **2 Independent Samples**.
2. Move the names of the relevant dependent variables to the **Test Variable List**.

3. Click on the relevant grouping variable and move it to the **Grouping Variable** box.

4. Ensure that **Mann-Whitney U** is checked.

5. Click on **OK**.

For related samples, after clicking on **Nonparametric** we would click **2 Related Samples** and ensure that **Wilcoxon** is checked in the **Test Type** dialog box.

S-PLUS

Various distribution-free tests described in the text are available from the Statistics menu. The following steps access the relevant dialog boxes.

1. Click on **Statistics**; click on **Compare Samples**; then
   (a) Click on **One Sample**, and click on **Signed Rank Test** for the Wilcoxon's signed rank test dialog box.
   (b) Click on **Two Sample**, and click on **Wilcoxon Rank Test** for the Wilcoxon–Mann–Whitney test dialog box.
   (c) Click on **k Samples**, and click on **Kruskal-Wallis Rank Test** or **Friedman Rank Test** to get the dialog box for the Kruskal–Wallis one-way analysis procedure, or the Friedman procedure for repeated measures.

All these distribution free tests are also available by a particular function in the command language; relevant functions are *wilcox.test*, *kruskal.test*, and *friedman.test*. For example, to apply the Wilcoxon–Mann–Whitney test to the data in two vectors, *x1* and *x2*, the command would be

$$wilcox.test(x1,x2),$$

and to apply the Wilcoxon's signed rank test to paired data contained in two vectors of the same length, *y1* and *y2*, the command would be

$$wilcox.test(y1,y2,paired=T).$$

The density, cumulative probability, quantiles, and random generation for the distribution of the Wilcoxon–Mann–Whitney rank sum statistic are also readily available by using *dwilcox*, *pwilcox*, *qwilcox*, and *rwilcox*, and these can be very useful in particular applications.

The *outer* function is extremely helpful if distribution-free confidence intervals are required. For example, in the electrode example in the text, if the data for each electrode are stored in vectors, *electrode1* and *electrode2*, then the required sums of pairs of observations needed can be found as

$$diff < -electrode1 - electrode2,$$
$$wij < -outer(diff, diff, ``+")/2.$$

The *cor* function can be used to calculate a variety of correlation coefficients, and *cor.test* can be used for hypothesis testing of these coefficients.

The bootstrap resampling procedure is available from the Statistics menu as follows.

1. Click on **Statistics**, click on **Resample**, and click on **Bootstrap** to access the Bootstrap dialog box.
2. Select the relevant data set, enter the expression for the statistic to estimate, and click on the **Options** tag to alter the number of bootstrap samples from the default value of 1000, if required.

EXERCISES

8.1. The data in Table 8.15 were obtained in a study reported by Hollander and Wolfe (1999). A measure of depression was recorded for each patient on both the first and second visit after initiation of therapy. Use the Wilcoxon signed rank test to assess whether the depression scores have changed over the two visits and construct a confidence interval for the difference.

TABLE 8.15
Depression Scores

Patient	Visit 1	Visit 2
1	1.83	0.88
2	0.50	0.65
3	1.62	0.60
4	2.48	2.05
5	1.68	1.06
6	1.88	1.29
7	1.55	1.06
8	3.06	3.14
9	1.30	1.29

TABLE 8.16
Test Scores of Dizygous Twins

Pair i	Twin X_i	Twin Y_i
1	277	256
2	169	118
3	157	137
4	139	144
5	108	146
6	213	221
7	232	184
8	229	188
9	114	97
10	232	231
11	161	114
12	149	187
13	128	230

8.2. The data in Table 8.16 give the test scores of 13 dizygous (nonidentical) male twins (the data are taken from Hollander and Wolfe, 1999). Test the hypothesis of independence versus the alternative that the twins' scores are positively correlated.

8.3. Reanalyze the SO2 data given in Table 8.11 by using a permutational approach, taking as the test statistic the sum of the scores in the group that had received training.

8.4. Investigate how the bootstrap confidence intervals for both the mean and median differences in the WISC data change with the size of the bootstrap sample.

8.5. Use the bootstrap approach to produce an approximate 95% confidence interval for the ratio of the two population variances of the WISC data used in this chapter. Investigate how the confidence interval changes with the number of bootstrap samples used.

8.6. A therapist is interested in discovering whether family psychotherapy is of any value in alleviating the symptoms of asthma in children suffering from the disease. A total of eight families, each having a child with severe asthma, is selected for study. As the response variable, the therapist uses the number of trips to the emergency room of a hospital following an asthma attack in a 2-month period. The data shown in Table 8.17 give the values of this variable for the eight children both before psychotherapy and after psychotherapy. Use a

TABLE 8.17
Number of Visits to the Emergency Room
in a 2-Month Period

| Patient | Psychotherapy | |
	Before	After
1	1	1
2	4	2
3	3	2
4	8	5
5	3	2
6	1	2
7	3	2
8	4	3

TABLE 8.18
PEF and S:C Measurements for 10 Schizophrenic Patients

PEF	S:C
48	3.32
42	3.74
44	3.70
35	3.43
36	3.65
28	4.50
30	3.85
13	4.15
22	5.06
24	4.27

suitable distribution-free test to determine whether there has been any change in the two periods, and calculate a distribution-free confidence interval for the treatment effect.

8.7. The data in Table 8.18 show the value of an index known as the psychomotor expressiveness factor (PEF) and the ratio of striatum to cerebellum radioactivity concentration 2 hours after injection of a radioisotope (the ratio is known as S:C), for 10 schizophrenic patients. The data were collected in an investigation of the dopamine hypothesis of schizophrenia. Calculate the values of Pearson's product moment correlation, Kendall's tau, and Spearman's rank correlation for the two variables, and each case, test whether the population correlation is zero.

9

Analysis of Categorical Data I: Contingency Tables and the Chi-Square Test

9.1. INTRODUCTION

Categorical data occur frequently in the social and behavioral sciences, where information about marital status, sex, occupation, ethnicity, and so on is frequently of interest. In such cases the measurement scale consists of a set of categories. Two specific examples are as follows.

1. *Political philosophy*: liberal, moderate, conservative.
2. *Diagnostic test for Alzheimer's disease*: symptoms present, symptoms absent.

(The categories of a categorical variable should be *mutually exclusive*; i.e., one and only one category should apply to each subject, unlike, say, the set of categories of liberal, Christian, Republican.)

Many categorical scales have a natural ordering. Examples are attitude toward legalization of abortion (disapprove in all cases, approve only in certain cases, approve in all cases), response to a medical treatment (excellent, good, fair, poor) and diagnosis of whether a patient is mentally ill (certain, probable, unlikely, definitely not). Such ordinal variables have largely been dealt with in the previous chapter, and here we shall concentrate on categorical variables having unordered categories, the so-called *nominal variables*. Examples are religious affiliation (Catholic, Jewish, Protestant, other), mode of transportation to work (automobile, bicycle, bus, walk) and favorite type of music (classical, country, folk, jazz, rock). For nominal variables, the order of listing the categories is irrelevant and the statistical analysis should not depend on that ordering. (It could, of course, be argued that some of these examples *do* have an associated natural order; type of music, for example, might be listed in terms of its cultural content as country, folk, jazz, classical, and rock!)

In many cases, the researcher collecting categorical data is most interested in assessing how pairs of categorical variables are related, in particular whether they are *independent* of one another. Cross-classifications of pairs of categorical variables, that is, *two-dimensional contingency tables*, are commonly the starting point for such an investigation. Such tables and the associated chi-square test should be familiar to most readers from their introductory statistics course, but for those people whose memories of these topics have become a little faded, the next section will (it is hoped) act as a refresher. Subsequent sections will consider a number of topics dealing with two-dimensional contingency tables not usually encountered in an introductory statistics course.

9.2. THE TWO-DIMENSIONAL CONTINGENCY TABLE

A two-dimensional contingency table is formed from cross-classifying two categorical variables and recording how many members of the sample fall in each cell of the cross-classification. An example of a 5×2 contingency table is given in Table 9.1, and Table 9.2 shows a 3×3 contingency table. The main question asked

TABLE 9.1

Two-Dimensional Contingency Table: Psychiatric Patients by
Diagnosis and Whether Their Treatment Prescribed Drugs

Diagnosis	Drugs	No Drugs
Schizophrenic	105	8
Affective disorder	12	2
Neurosis	18	19
Personality disorder	47	52
Special symptoms	0	13

TABLE 9.2

Incidence of Cerebral Tumors

Site	Type A	B	C	Total
I	23	9	6	38
II	21	4	3	28
III	34	24	17	75
Total	78	37	26	141

Note. Sites: I, frontal lobes; II, temporal lobes; III, other cerebral areas. Types: A, benign tumors; B, malignant tumors; C, other cerebral tumors.

TABLE 9.3

Estimated Expected Values Under the Hypothesis of Independence for the Diagnosis and Drugs Data in Table 9.1

Diagnosis	Drugs	No Drugs
Schizophrenic	74.51	38.49
Affective disorder	9.23	4.77
Neurosis	24.40	12.60
Personality disorder	65.28	33.72
Special symptoms	8.57	4.43

about such tables is whether the two variables forming the table are independent or not. The question is answered by use of the familiar *chi-squared test*; details are given in Display 9.1. (Contingency tables formed from more than two variables will be discussed in the next chapter.)

Applying the test described in Display 9.1 to the data in Table 9.1 gives the estimated expected values shown in Table 9.3 and a chi-square value of 84.19 with 4 degrees of freedom. The associated p value is very small. Clearly, diagnosis and treatment with drugs are not independent. For Table 9.2 the chi-square statistic takes the value 7.84, which with 4 degrees of freedom leads to a p value of .098. Here there is no evidence against the independence of site and type of tumor.

Display 9.1

Testing for Independence in an $r \times c$ Contingency Table

---

- Suppose a sample of n individuals have been cross-classified with respect to two categorical variables, one with r categories and one with c categories, to form an $r \times c$ two-dimensional contingency table.
- The general form of a two-dimensional contingency table is as follows.

		Variable 2				
		1	2	$\cdots$	c	Total
	1	n_{11}	n_{12}	$\cdots$	n_{1c}	$n_{1.}$
	2	n_{21}	n_{22}	$\cdots$	n_{2c}	$n_{2.}$
Variable 1	$\vdots$	$\vdots$	$\vdots$	$\vdots$	$\vdots$	$\vdots$
	r	n_{r1}	n_{r2}	$\cdots$	n_{rc}	$n_{r.}$
	Total	$n_{.1}$	$n_{.2}$	$\cdots$	$n_{.c}$	n

- Here n_{ij} represents the number of observations in the ijth cell of the table, $n_{i.}$ represents the total number of observations in the ith row of the table and $n_{.j}$ represents the total number of observations in the jth column of the table—both $n_{i.}$ and $n_{.j}$ are usually termed *marginal totals*.
- The null hypothesis to be tested is that the two variables are independent. This hypothesis can be formulated more formally as

$$H_0 : p_{ij} = p_{i.} \times p_{.j},$$

where in the population from which the n observations have been sampled, p_{ij} is the probability of an observation being in the ijth cell, $p_{i.}$ is the probability of being in the ith category of the row variable, and $p_{.j}$ is the probability of being in the jth category of the column variable. (The hypothesis is just a reflection of that elementary rule of probability that for two independent events A and B, the probability of A *and* B is simply the product of the probabilities of A and of B.)
- In a sample of n individuals we would, if the variables are independent, expect $np_{i.}p_{.j}$ individuals in the ijth cell.
- Using the obvious estimators for $p_{i.}$ and $p_{.j}$, we can estimate this expected value to give

$$E_{ij} = n \times \frac{n_{i.}}{n} \times \frac{n_{.j}}{n} = \frac{n_{i.} \times n_{.j}}{n}.$$

- The observed and expected values under independence are then compared by using the familiar chi-squared statistic

$$X^2 = \sum_{i=1}^{r} \sum_{j=1}^{c} \frac{(n_{ij} - E_{ij})^2}{E_{ij}}.$$

- If independence holds then the X^2 statistic has approximately a chi-squared distribution with $(r-1)(c-1)$ degrees of freedom. This allows p values to be assigned.
- The possible problems caused by the use of the chi-square distribution to approximate the true null distribution of X^2 is taken up in the text.

TABLE 9.4
Some Examples of 2 × 2 Contingency Tables

1. Classification of Psychiatric Patients by Sex and Diagnosis

	Sex		
Diagnosis	*Male*	*Female*	*Total*
Schizophrenia	43	32	75
Other	15	32	47
Total	58	64	122

2. Data from Pugh (1983) Involving How Juries Come to Decisions in Rape Cases[a]

	Verdict		
Fault	*Guilty*	*Not Guilty*	*Total*
Not alleged	153	24	177
Alleged	105	76	181
Total	258	100	358

3. Incidence of Suicidal Feelings in Psychotic and Neurotic Patients

	Type of Patient		
Suicidal Feelings	*Psychotic*	*Neurotic*	*Total*
Yes	2	6	8
No	18	14	32
Total	20	20	40

[a]Here the verdict is classified against whether the defense alleged that the victim was somehow partially at fault for the rape.

The simplest form of two-dimensional contingency table is obtained when a sample of observations is cross-classified according to the values taken by two *dichotomous* variables, that is, categorical variables with only two categories. Several examples of such tables are shown in Table 9.4. The general form of such a 2 × 2 contingency table, the special form of the chi-square test for such tables, and the construction of a useful confidence interval associated with 2 × 2 tables are described in Display 9.2.

The results of applying the chi-squared test to the data sets in Table 9.4 and the derived confidence intervals for differences in proportions are shown in Table 9.5.

Display 9.2

2 × 2 Contingency Tables

- The general form of a 2 × 2 contingency table is as follows.

		Variable 2		
		Category 1	Category 2	Total
	Category 1	a	b	$a+b$
Variable 1				
	Category 2	c	d	$c+d$
	Total	$a+c$	$b+d$	$n = a+b+c+d$

- The chi-squared statistic used in testing for independence can now be written in simplified form as

$$X^2 = \frac{n(ad - bc)^2}{(a+b)(c+d)(a+c)(b+d)}.$$

- For a 2 × 2 table the statistic has a single degree of freedom.
- For this type of contingency table, independence implies that the probability of being in category 1 of variable 1 and category 1 of variable 2 (p_1) is equal to the probability of being in category 1 of variable 1 and category 2 of variable 2 (p_2).
- Estimates of these two probabilities are given by

$$\hat{p}_1 = \frac{a}{a+c}, \qquad \hat{p}_2 = \frac{b}{b+d}.$$

- The standard error of the difference of the two estimates is given by

$$SE(\hat{p}_1 - \hat{p}_2) = \sqrt{\frac{\hat{p}_1(1 - \hat{p}_1)}{a+c} + \frac{\hat{p}_2(1 - \hat{p}_2)}{b+d}}.$$

- This can be used to find a confidence interval for the difference in the two probabilities in the usual way.

TABLE 9.5

Results of Analyzing the Three 2 × 2 Contingency Tables in Table 9.4

1. For the schizophrenia and gender data, $X^2 = 7.49$ with an associated p value of .0062. The 95% confidence interval for the difference in the probability of being diagnosed schizophrenic for men and for women is (0.01, 0.41).
2. For the rape data, $X^2 = 35.93$ with an associated p value that is very small. The 95% confidence interval for the difference in the probability of being found guilty or not guilty when the defense does not suggest that the rape is partially the fault of the victim is (0.25, 0.46). . . . i.e., defendent more likely to be found guilty.
3. For the suicidal feelings data, $X^2 = 2.5$ with an associated p value of .11. The 95% confidence interval for the difference in the probability of having suicidal feelings in the two diagnostic categories is (−0.44, 0.04).

The results indicate the following.

1. That sex and psychiatric diagnosis are associated in the sense that a higher proportion of men than women are diagnosed as being schizophrenic.
2. That the verdict in a rape case is associated with whether or not the defense allege that the rape was partially the fault of the victim.
3. Diagnosis and suicidal feelings are not associated.

Three further topics that should be mentioned in connection with 2×2 contingency tables are

- *Yates's continuity correction*—see Display 9.3;
- *Fisher's exact test*—see Display 9.4;
- *McNemar's test for matched samples*—see Display 9.5.

We can illustrate the use of Fisher's exact test on the data on suicidal feelings in Table 9.4 because this has some small expected values (see Section 9.4 for more comments). The p-value form applying the test is .235, indicating that diagnosis and suicidal feelings are not associated.

To illustrate McNemar's test, we use the data shown in Table 9.6. For these data the test statistic takes the value 1.29, which is clearly not significant, and we can conclude that depersonalization is not associated with prognosis where endogenous depressed patients are concerned.

Display 9.3
Yates's Continuity Correction

- In the derivation of the null distribution of the X^2 statistic, a *continuous* probability distribution, namely the chi-square distribution, is being used as an approximation to the *discrete* probability distribution of observed frequencies, namely the *multinomial distribution* (see glossary in Appendix A).
- To improve this approximation, Yates (1934) suggested a correction to the test statistic that involves subtracting 0.5 from the positive discrepancies (observed–expected), and adding 0.5 to the negative discrepancies, before these values are squared in the calculation of X^2.
- The correction may be incorporated into the formula for X^2 for the 2×2 table given in Display 9.2 to become

$$X^2 = \frac{n(|ad - bc| - 0.5n)^2}{(a + b)(c + d)(a + c)(b + d)}.$$

- This is now known as the chi-square value corrected for continuity.
- Although Yates's correction is still widely used, it is really no longer necessary because of the routine availability of the exact methods described in Section 9.4.

Display 9.4

Fisher's Exact Test for 2×2 Contingency Tables

- Fisher's exact test for a 2×2 contingency table does not use the chi-square approximation at all. Instead the exact probability distribution of the observed frequencies is used.
- For fixed marginal totals, the required distribution is what is known as a *hypergeometric distribution*. Assuming that the two variables forming the table are independent, the probability of obtaining any particular arrangement of the frequencies $a, b, c,$ and d when the marginal totals are as given is

$$\Pr(a, b, c, d) = \frac{(a+b)!(a+c)!(c+d)!(b+d)!}{a!b!c!d!n!},$$

  where $a!$—read a factorial—is the shorthand method of writing the product of a and all the integers less than it. (By definition, $0!$ is one.)
- Fisher's test uses this formula to find the probability of the observed arrangement of frequencies, *and* of every other arrangement giving as much or more evidence of an association between the two variables, keeping in mind that the marginal totals are regarded as fixed.
- The sum of the probabilities of all such tables is the relevant p value for Fisher's test.

Display 9.5

McNemar's Test for Paired Samples

- When the samples to be compared in a 2×2 contingency table are matched in some way, for example, the same subjects observed on two occasions, then the appropriate test becomes one from McNemar (1962).
- For a matched data set, the 2×2 table takes the following form.

		Sample 1	
		A present	A absent
	A present	a	b
Sample 2			
	A absent	c	d

  where A is the characteristic assessed in the pairs of observation making up the matched samples. Now a represents the number of *pairs* of observations that both have A, and so on.
- To test whether the probability of having A differs in the matched populations, the relevant test statistic is

$$X^2 = \frac{(b-c)^2}{b+c},$$

  which, if there is no difference, has a chi-squared distribution with a single degree of freedom.

TABLE 9.6
Recovery of 23 Pairs of Depressed Patients

| | | Depersonalized Patients | | |
		Recovered	Not Recovered	Total
Patients Not Depersonalized	Recovered	14	5	19
	Not recovered	2	2	4
	Total	16	7	23

9.3. BEYOND THE CHI-SQUARE TEST: FURTHER EXPLORATION OF CONTINGENCY TABLES BY USING RESIDUALS AND CORRESPONDENCE ANALYSIS

A statistical significance test is, as implied in Chapter 1, often a crude and blunt instrument. This is particularly true in the case of the chi-square test for independence in the analysis of contingency tables, and after a significant value of the test statistic is found, it is usually advisable to investigate in more detail *why* the null hypothesis of independence fails to fit. Here we shall look at two approaches, the first involving suitably chosen residuals and the second that attempts to represent the association in a contingency table graphically.

9.3.1. The Use of Residuals in the Analysis of Contingency Tables

After a significant chi-squared statistic is found and independence is rejected for a two-dimensional contingency table, it is usually informative to try to identify the cells of the table responsible, or most responsible, for the lack of independence. It might be thought that this can be done relatively simply by looking at the deviations of the observed counts in each cell of the table from the estimated expected values under independence, that is, by examination of the residuals:

$$r_{ij} = n_{ij} - E_{ij}, \quad i = 1, \ldots, r, \quad j = 1, \ldots, c. \quad (9.1)$$

This would, however, be very unsatisfactory because a difference of fixed size is clearly more important for smaller samples. A more satisfactory way of defining residuals for a contingency table might be to take

$$e_{ij} = (n_{ij} - E_{ij})/\sqrt{E_{ij}}. \tag{9.2}$$

These terms are usually known as *standardized residuals* and are such that the chi-squared test statistic is given by

$$X^2 = \sum_{i=1}^{r}\sum_{j=1}^{c} e_{ij}^2. \tag{9.3}$$

It is tempting to think that these residuals might be judged as standard normal variables, with values outside say $(-1.96, 1.96)$ indicating cells that depart significantly from independence. Unfortunately, it can be shown that the variance of e_{ij} is always less than or equal to one, and in some cases considerably less than one. Consequently the use of standardized residuals for a detailed examination of a contingency table may often give a conservative reading as to which cells independence does not apply.

At the cost of some extra calculation, a more useful analysis can be achieved by using what are known as *adjusted residuals*, d_{ij}, as suggested by Haberman (1973). These are defined as follows:

$$d_{ij} = \frac{e_{ij}}{[(1 - n_{i.}/n)(1 - n_{.j}/n)]}. \tag{9.4}$$

When the variables forming the contingency table are independent, the adjusted residuals are approximately normally distributed with mean zero and standard deviation one. Consequently, values outside $(-1.96, 1.96)$ indicate those cells departing most from independence.

Table 9.7 shows the adjusted residuals for the data in Table 9.1. Here the values of the adjusted variables demonstrate that *all* cells in the table contribute to the

TABLE 9.7
Adjusted Residuals for the Data in Table 9.1

Diagnosis	Drugs	No Drugs
Schizophrenic	151.56	−78.28
Affective disorder	8.56	−4.42
Neurosis	−21.69	11.21
Personality disorder	−83.70	43.23
Special symptoms	−26.41	13.64

departure from independence of diagnosis and treatments with drugs. (See Exercise 9.6 for further work with residuals.)

9.3.2. Correspondence Analysis

Correspondence analysis attempts to display graphically the relationship between the variables forming a contingency table by deriving a set of coordinate values representing the row and column categories of the table. The correspondence analysis coordinates are analogous to those derived from a *principal components analysis* of continuous, multivariate data (see Everitt and Dunn, 2001), except that they are derived by partitioning the chi-squared statistic for the table, rather than a variance. A brief nontechnical account of correspondence analysis is given in Display 9.6; a full account of the technique is available in Greenacre (1984).

As a first example of using a correspondence analysis, it will be applied to the data shown in Table 9.8. The results of the analysis are shown in Table 9.9 and the resulting correspondence analysis diagram in Figure 9.1. The pattern is as would

---

Display 9.6

Correspondence Analysis

- Correspondence analysis attempts to display graphically the relationship between the two variables forming a contingency table by deriving a sets of coordinates representing the row and columns categories of the table.
- The coordinates are derived from a procedure known as *singular value decomposition* (see Everitt and Dunn, 2001, for details) applied to the matrix **E**, the elements of which are

$$e_{ij} = \frac{n_{ij} - E_{ij}}{\sqrt{E_{ij}}},$$

where the terms are as defined in Display 9.1.
- Application of the method leads to two sets of coordinates, one set representing the row categories and the other representing the column categories. In general, the first two coordinate values for each category are the most important, because they can be used to plot the categories as a scatterplot.
- For a two-dimensional representation, the row category coordinates can be represented as u_{ik}, $i = 1, \ldots, r$, and $k = 1, 2$, and the column category coordinates as v_{jk}, $j = 1, \ldots, c$, and $k = 1, 2$.
- A cell for which the row and column coordinates are both large and of the same sign is one that has a larger observed value than that expected under independence. A cell for which the row and column coordinates are both large but of opposite signs is one in which the observed frequency is lower than expected under independence. Small coordinate values for the row and column categories of a cell indicate that the observed and expected values are not too dissimilar.
- Correspondence analysis diagrams can be very helpful when contingency tables are dealt with, but some experience is needed in interpreting the diagrams.

TABLE 9.8

Cross-Classification of Eye Color and Hair Color

	Hair Color				
Eye Color	Fair	Red	Medium	Dark	Black
Light	688	116	584	188	4
Blue	326	38	241	110	3
Medium	343	84	909	412	26
Dark	98	48	403	681	81

TABLE 9.9

Derived Coordinates from Correspondence Analysis for
Hair Color–Eye Color Data

Category	Coord 1	Coord 2
Eye light (EL)	0.44	0.09
Eye blue (EB)	0.40	0.17
Eye medium (EM)	−0.04	−0.24
Eye dark (ED)	−0.70	0.14
Hair fair (hf)	0.54	0.17
Hair red (hr)	0.23	0.05
Hair medium (hm)	0.04	−0.21
Hair dark (hd)	−0.59	0.11
Hair brown (hb)	−1.08	0.27

be expected with, for example, fair hair being associated with blue and light eyes, and so on.

The second example of a correspondence analysis will involve the data shown in Table 9.10, collected in a survey in which people in the UK were asked about which of a number of characteristics could be applied to people in the UK and in other European countries. Here the two-dimensional correspondence diagram is shown in Figure 9.2. It appears that the respondents judged, for example, the French to be stylish and sexy, the Germans efficient and the British boring—well, they do say you can prove anything with statistics!

TABLE 9.10
What do People in the UK Think about Themselves and Their Partners in the
European Community?

Nationality	\	\	\	\	\	Characteristic	\	\	\	\	\	\	\
	1 (sty)	2 (arr)	3 (sex)	4 (dev)	5 (eas)	6 (gre)	7 (cow)	8 (bor)	9 (eff)	10 (laz)	11 (har)	12 (cle)	13 (cou)
French (F)	37	29	21	19	10	10	8	8	6	6	5	2	1
Spanish (S)	7	14	8	9	27	7	3	7	3	23	12	1	3
Italian (I)	30	12	19	10	20	7	12	6	5	13	10	1	2
British (B)	9	14	4	6	27	12	2	13	26	16	29	6	25
Irish (Ir)	1	7	1	16	30	3	10	9	5	11	22	2	27
Dutch (D)	5	4	2	2	15	2	0	13	24	1	28	4	6
German (G)	4	48	1	12	3	9	2	11	41	1	38	8	8

Note. 1, stylish; 2, arrogant; 3, sexy; 4, devious; 5, easygoing; 6, greedy; 7, cowardly; 8, boring;
9, efficient; 10, lazy; 11, hardworking; 12, clever; 13, courageous.

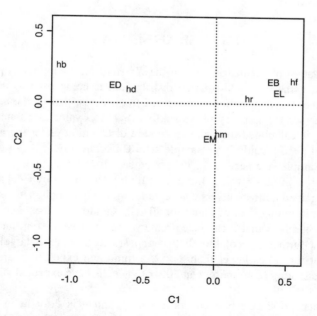

FIG. 9.1. Correspondence analysis diagram for the hair color–Eye
color data in Table 9.8.

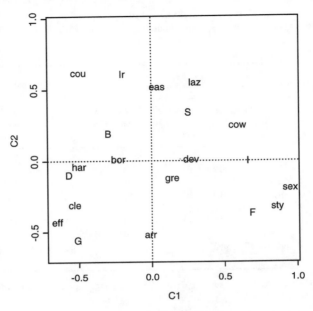

FIG. 9.2. Correspondence analysis diagram for the data in
Table 9.10.

9.4. SPARSE DATA

In the derivation of the distribution of the X^2 statistic, a continuous probability
distribution, namely the chi-squared distribution, is being used as an approximation to the true distribution of observed frequencies, that is, the *multinomial
distribution* (see the glossary in Appendix A). The *p* value associated with the
X^2 statistic is calculated from the chi-square distribution under the assumption
that there is a sufficiently large sample size. Unfortunately, it is not easy to describe the sample size needed for the chi-square distribution to approximate the
exact distribution of X^2 well. One rule of thumb suggested by Cochran (1954),
which has gained almost universal acceptance among psychologists (and others),
is that the minimum expected count for all cells should be at least five. The problem with this rule is that it can be extremely conservative, and, in fact, Cochran
also gave a further rule of thumb that appears to have been largely ignored,
namely that for tables larger than 2×2, a minimum expected count of one is
permissible as long as no more than 20% of the cells have expected values below
five.

In the end no simple rule covers all cases, and it is difficult to identify, *a
priori*, whether or not a given data set is likely to suffer from the usual asymptotic

TABLE 9.11
Firefighters' Entrance Exam Results

| Test Result | Ethnic Group of Entrant | | | | Total |
	White	Black	Asian	Hispanic	
Pass	5	2	2	0	9
No show	0	1	0	1	2
Fail	0	2	3	4	9
Total	5	5	5	5	20

TABLE 9.12
Reference Set for the Firefighter Example

x_{11}	x_{12}	x_{13}	x_{14}	9
x_{21}	x_{22}	x_{23}	x_{24}	2
x_{31}	x_{32}	x_{33}	x_{34}	9
5	5	5	5	20

inference. One solution that is now available is to compute *exact p* values by using a permutational approach of the kind encountered in the previous chapter. The main idea in evaluating exact p values is to evaluate the observed table, relative to a reference set of other tables of the same size that are like it in every possible respect, except in terms of reasonableness under the null hypothesis. The approach will become clearer if we use the specific example of Table 9.11, which summarizes the results of a firefighters' entrance exam. The reference set of tables for this example is shown in Table 9.12. The X^2 statistic for Table 9.11 is 11.56; but here the number of cells with expected value less than 5 is 12, that is, all cells in the table! The exact p value is then obtained by identifying all tables in the reference set whose X^2 values equal or exceed the observed statistic, and summing their probabilities, which under the null hypothesis of independence are found from the hypergeometric distribution formula (see Display 9.4). For example, Table 9.13(a) is a member of the reference set and has a value for X^2 of 14.67. Its exact probability is .00108, and because its X^2 value is more extreme than the observed value, it will contribute to the exact p value. Again, Table 9.13(b)

TABLE 9.13

Two Tables in the Reference Set for the Firefighter Data

(a)

5	2	2	0	9
0	0	0	2	2
0	3	3	3	9
5	5	5	5	20

(b)

4	3	2	0	9
1	0	0	1	2
0	2	3	4	9
5	5	5	5	20

Note. $X^2 = 14.67$, and the exact probability is .00108, for (a). $X^2 = 9.778$ (not larger that the observed X^2 value) and so it does not contribute to the exact p value, for (b).

TABLE 9.14

Reference Set for a Hypothetical 6×6 Contingency Table

x_{11}	x_{12}	x_{13}	x_{14}	x_{15}	x_{16}	7
x_{21}	x_{22}	x_{23}	x_{24}	x_{25}	x_{26}	7
x_{31}	x_{32}	x_{33}	x_{34}	x_{35}	x_{36}	12
x_{41}	x_{42}	x_{43}	x_{44}	x_{45}	x_{46}	4
x_{51}	x_{52}	x_{53}	x_{54}	x_{55}	x_{56}	4
4	5	6	5	7	7	34

is also a member of the reference set. Its X^2 value is 9.778, which is less than that of the observed table and so its probability does not contribute to the exact p value. The exact p value calculated in this way is .0398. The asymptotic p value associated with the observed X^2 value is .07265. Here the exact approach leads to a different conclusion, namely that the test result is not independent of race.

The real problem in calculating exact p values for contingency tables is computational. For example, the number of tables in the reference set for Table 9.14 is 1.6 billion! Fortunately, methods and associated software are now available that make this approach a practical possibility (see StatXact, Cytel Software Corporation; pralay@cytel.com.).

TABLE 9.15
Calculation of the Odds Ratio and Its Confidence Interval for the Schizophrenia
and Gender Data

- The estimate of the odds ratio is

$$\hat{\psi} = (43 \times 52)/(32 \times 15) = 4.66$$

- The odds in favor of being disgnosed schizophrenic among males is nearly five times the corresponding odds for females.
- The estimated variance of the logarithm of the estimated odds ratio is

$$1/4 + 1/52 + 1/15 + 1/32 = 0.1404$$

- An approximate confidence for $\log \psi$ is

$$\log(4.66) \pm 1.96 \times \sqrt{0.1404} = (0.80, 2.29).$$

- Consequently, the required confidence interval for ψ is

$$[\exp(0.80), \exp(2.29)] = [2.24, 9.86].$$

9.5. THE ODDS RATIO

Defining the independence of the two variables forming a contingency table is relatively straightforward, but measuring the degree of dependence is not so clear cut, and many measures have been proposed (Everitt, 1992). In a 2×2 table one possible measure might be thought to be the difference between the estimates of the two probabilities of interest, as illustrated in Display 9.2. An alternative, and in many respects a more acceptable, measure is the *odds ratio*, which is explained in Display 9.7. This statistic is of considerable practical importance in the application of both *log-linear models* and *logistic regression* (see Chapter 10). The calculation of the odds ratio and its standard error for the schizophrenia and gender data are outlined in Table 9.15. The odds of being diagnosed schizophrenic in males is between approximately 2 and 10 times the odds in females.

9.6. MEASURING AGREEMENT
FOR CATEGORICAL VARIABLES:
THE KAPPA STATISTIC

It is often required to measure how well two observers agree on the use of a categorical scale. The most commonly used index of such agreement is the *kappa coefficient*, first suggested by Cohen (1960). The data shown in Table 9.16 will be used to illustrate the use of this index. These data are taken from a study (Westland and Kurland, 1953) in which two neurologists independently classified

Display 9.7
The Odds Ratio

- The general 2×2 contingency table met in Display 9.2 can be summarized in terms of the probabilities of an observation being in each of the four cells of the table.

	Variable 2	
Variable 1	Category 1	Category 2
Category 1	p_{11}	p_{12}
Category 2	p_{21}	p_{22}

- The ratios p_{11}/p_{12} and p_{21}/p_{22} are know as *odds*. The first is the odds of being in category 1 of variable 1 for the two categories of variable 2. The second is the corresponding odds for category 2 of variable 1. (Odds will be familiar to those readers who like the occasional flutter on the Derby—Epsom or Kentucky!)
- A possible measure for the degree of dependence of the two variables forming a 2×2 contingency table is the so-called *odds ratio*, given by

$$\psi = \frac{p_{11}}{p_{12}} \bigg/ \frac{p_{21}}{p_{22}} = \frac{p_{11} p_{22}}{p_{21} p_{12}}.$$

- Note that ψ may take any value between zero and infinity, with a value of 1 corresponding to independence (why?).
- The odds ratio, ψ, has a number of desirable properties for representing dependence among categorical variables that other competing measures do not have. These properties include the following.
  1. ψ remains unchanged if the rows and columns are interchanged.
  2. If the levels of a variable are changed (i.e., listing category 2 before category 1), ψ becomes $1/\psi$.
  3. Multiplying either row by a constant or either column by a constant leaves ψ unchanged.
- The odds ratio is estimated from the four frequencies in an observed 2×2 contingency table (see Display 9.2) as

$$\hat{\psi} = ad/bc.$$

- Confidence intervals for ψ can be determined relatively simply by using the following estimator of the variance of $\log \psi$:

$$\text{vâr}(\log \psi) = 1/a + 1/b + 1/c + 1/d.$$

- An approximate 95% confidence inteval for $\log \psi$ is given by

$$\log \hat{\psi} \pm 1.96 \times \sqrt{\text{vâr}(\log \hat{\psi})}.$$

- If the limits of the confidence interval for $\log \psi$ obtained in this way are ψ_L, ψ_U, then the corresponding confidence interval for ψ is simply $\exp(\psi_L)$, $\exp(\psi_U)$.

TABLE 9.16
Diagnosis of Multiple Sclerosis by Two Neurologists

	A	B	C	D	Total
A	38	5	0	1	44
B	33	11	3	0	47
C	10	14	5	6	35
D	3	7	3	10	23
Total	84	37	11	17	149

Source. Westland and Kurland (1953).

149 patients into four classes: (1) A, certainly suffering from multiple sclerosis; (2) B, probably suffering from multiple sclerosis; (3) C, possibly suffering from multiple sclerosis; and (4) D, doubtful, unlikely, and definitely not suffering from multiple sclerosis.

One intuitively reasonable index of agreement for the two raters is the proportion P_0 of patients that they classify into the same category; for the data in Table 9.16,

$$P_0 = (38 + 11 + 5 + 10)/149 = 0.429. \qquad (9.5)$$

Such a measure has the virtue of simplicity and it is readily understood. However, despite such advantages (and intuition), P_0 is *not* an adequate index of the agreement between the two raters. The problem is that P_0 makes no allowance for agreement between raters that might be attributed to chance. To explain, consider the two sets of data in Table 9.17. In both, the two observers are measured as achieving 66% agreement if P_0 is calculated. Suppose, however, that each observer is simply allocating subjects at random to the three categories in accordance with their *marginal rates* for the three categories. For example, observer A in the first data set would simply allocate 10% of subjects to category 1, 80% to category 2, and the remaining 10% to category 3, totally disregarding the suitability of a category for a subject. Observer B proceeds likewise.

Even such a cavalier rating procedure used by the two observers would lead to some agreement and a corresponding nonzero value of P_0. This *chance agreement*, P_c, can be calculated simply from the marginal rates of each observer. For example, for the first data set in Table 9.17, P_c is calculated as follows.

1. Category 1: the number of chance agreements to be expected is

$$100 \times 10/100 \times 10/100 = 1. \qquad (9.6)$$

TABLE 9.17
Two Hypothetical Data Sets, Each of which Shows 66%
Agreement Between the Two Observers

Observer A		Observer B		Total
Data Set 1				
	1	2	3	
1	1	8	1	10
2	8	64	8	80
3	1	8	1	10
Total	10	80	10	100
Data Set 2				
1	24	13	3	40
2	5	20	5	30
3	1	7	22	30
Total	30	40	30	100

(Remember how "expected" values are calculated in contingency tables? See Display 9.1.)

2. Category 2: the number of chance agreements to be expected is

$$100 \times 80/100 \times 80/100 = 64. \tag{9.7}$$

3. Category 3: the number of chance agreements to be expected is

$$100 \times 10/100 \times 10/100 = 1. \tag{9.8}$$

Consequently, P_c is given by

$$P_c = \frac{1}{100}[1 + 64 + 1] = 0.66. \tag{9.9}$$

Therefore, in this particular table, *all* the observed agreement might simply be due to chance. However, repeating the calculation on the second set of data in Table 9.17 gives $P_c = 0.33$, which is considerably lower than the observed agreement.

A number of authors have expressed opinions on the need to incorporate chance agreement into the assessment of interobserver reliability. The clearest statement in favor of such a correction has been made by Fleiss (1975), who suggested an index that is the ratio of the difference between observed and chance agreement to the maximum possible excess of observed over chance agreement, that is, $1 - P_c$.

This leads to what has become known as the kappa statistic:

$$\kappa = (P_0 - P_c)/(1 - P_c). \tag{9.10}$$

If there is complete agreement between the two raters so that all the off-diagonal cells of the table are empty, $\kappa = 1$. If observed agreement is greater than chance, $\kappa > 0$. If the observed agreement is equal to chance, $\kappa = 0$. Finally, in the unlikely event of the observed agreement being less than chance, $\kappa > 0$, with its minimum value depending on the marginal distributions of the two raters.

The chance agreement for the multiple sclerosis data is given by

$$P_c = \frac{1}{149}(44/149 \times 84 + 47/149 \times 37 + 35/149 \times 11$$
$$+ 23/149 \times 17) = 0.2797. \tag{9.11}$$

Consequently, for the multiple sclerosis data,

$$\kappa = (0.429 - 0.280)/(1 - 0.280) = 0.208. \tag{9.12}$$

This calculated value of κ is an estimate of the corresponding population value and, like all such estimates, has to be accompanied by some measure of its variance so that a confidence interval can be constructed. The variance of an observed value of κ has been derived under a number of different assumptions by several authors, including Everitt (1968) and Fleiss, Cohen, and Everitt (1969). The formula for the large-sample variance of κ is rather unpleasant, but for those with a strong stomach it is reproduced in Display 9.8. Its value for the multiple sclerosis data is 0.002485, which leads to an approximate 95% confidence interval of (0.108, 0.308). Thus there is some evidence that the agreement between the two raters in this example is greater than chance; otherwise the confidence interval would have included the value zero. However, what constitutes "good" agreement? Some arbitrary benchmarks for the evaluation of observed κ values have been given by

Display 9.8
Large Sample Variance of Kappa

$$\mathrm{var}(\kappa) = \frac{1}{n(1 - P_c)^4}\left\{ \sum_{i=1}^{r} p_{ii}[(1 - P_c) - (p_{.i} + p_{.j})(1 - P_o)]^2 \right.$$
$$\left. +(1 - P_o)^2 \sum_{\substack{i=1 \\ i \neq j}}^{r} \sum_{j=1}^{r} p_{ij}(p_{.i} + p_{j.})^2 - (P_o P_c - 2P_c + P_o)^2 \right\},$$

where p_{ij} is the proportion of observations in the ijth cell of the table of counts of agreements and disagreements for the two observers, $p_{i.}$ and $p_{.j}$ are the row and column marginal proportions, and r is the number of rows and columns in the table.

Landis and Koch (1977). They are as follows.

kappa	Strength of Agreement
0.00	Poor
0.01–0.20	Slight
0.21–0.40	Fair
0.41–0.60	Moderate
0.61–0.80	Substantial
0.81–1.00	Perfect

Of course, any series of standards such as these are necessarily subjective. Nevertheless, they may be helpful in the informal evaluation of a series of κ values, although replacing numerical values with rather poorly defined English phrases may not be to everybody's taste. In fact, there is no simple answer to the original question concerning what constitutes good agreement. Suppose, for example, that two examiners rating examination candidates as pass or fail had $\kappa = 0.54$ (in the moderate range according to Landis and Koch). Would the people taking the examination be satisfied by this value? This is unlikely, particularly if future candidates are going to be assessed by one of the examiners but not both. If this were the case, sources of disagreement should be searched for and rectified; only then might one have sufficient confidence in the assessment of a lone examiner.

The concept of a chance corrected measure of agreement can be extended to situations involving more than two observers; for details, see Fleiss and Cuzick (1979) and Schouten (1985). A weighted version of κ is also possible, with weights reflecting differences in the seriousness of disagreements. For example, in the multiple sclerosis data a disagreement involving one rater classifying a patient as A and the other rater classifying the same patient as D would be very serious and would be given a high weight. An example of the calculation of weighted κ and some comments about choice of weights are given by Dunn (1989).

9.7. SUMMARY

1. Categorical data occur frequently in psychological studies.
2. The chi-square statistic can be used to assess the independence or otherwise of two categorical variables.
3. When a significant chi-square value for a contingency table has been obtained, the reasons for the departure from independence have to be examined in more detail by the use of adjusted residuals, correspondence analysis, or both.
4. Sparse data can be a problem when the 2×2 statistic is used. This is a problem that can be overcome by computing exact p values.

5. The odds ratio is an extremely useful measure of association.
6. The kappa statistic can be used to quantify the agreement between two observers applying a categorical scale.

COMPUTER HINTS

SPSS

The most common test used on categorical data, the chi-square test for independence in a two-dimensional contingency table, can be applied as follows.

1. Click on **Statistics**, click on **Summarize**, and then click on **Crosstabs**.
2. Specify the row variable in the **Rows** box and the column variable in the **Columns** box.
3. Now click on **Statistics** and select **Chi-square**.
4. To get observed and expected values, click on **Continue**, click on **Cells**, and then click on **Expected** and **Total** and ensure that **Observed** is also checked.
5. Click on **Continue** and then **OK**.

S-PLUS

Various tests described in the text can be accessed from the Statistics menu as follows.

1. Click on **Statistics**, click on **Compare Samples**, and click on **Counts and Proportions**. Then,
   (a) Click on **Fisher's exact test** for the Fisher test dialog box,
   (b) Click on **McNemar's test** for the McNemar test dialog box,
   (c) Click on **Chi-square test** for the chi-square test of independence.

These tests can also be used in the command line approach, with the relevant functions being *chisq.test*, *fisher.test*, and *mcnemar.test*. For example, for a contingency table in which the frequencies were stored in a matrix X, the chi-square test of independence could be applied with the command

$$chisq.test(X).$$

If X is a 2×2 table, then Yates's correction would be applied by default. To stop the correction being applied would require

$$chisq.test(X, correct = F).$$

EXERCISES

9.1. Table 9.18 shows a cross-classification of gender and belief in the afterlife for a sample of Americans. Test whether the two classifications are independent, and construct a confidence interval for the difference in the proportion of men and proportion of women who believe in the afterlife.

9.2. Table 9.19 shows data presented in the case of *US versus Lansdowne Swimming Club*. Analyzed in the usual way by the Pearson chi-squared statistic, the results are statistically significant. The government, however, lost the case, because of applying the statistic when the expected count in two cells was less than five—the defendant argued that the software used to analyze the data printed out a warning that the chi-square test might not be valid. Use an exact test to settle the question.

9.3. Lundberg (1940) presented the results of an experiment to answer the question, What is the degree of agreement in commonsense judgments of socio-economic status by two persons who are themselves of radically different status?

TABLE 9.18
Gender and Belief in the Afterlife

Gender	Belief in the Afterlife	
	Yes	No
Female	435	147
Male	375	134

TABLE 9.19
Application for Membership at the Lansdowne Swimming Club

Parameter	Black Applicants	White Applicants
Accepted for membership	1	379
Rejected for membership	5	0
Total applicants	6	379

TABLE 9.20
Janitor's and Banker's Ratings of the Socioeconomic Status of 196 Families on a
Six-Point Scale

Banker's Ratings	Janitor's Ratings					
	1	*2*	*3*	*4*	*5*	*6*
6	0	0	0	0	0	0
5	0	0	0	6	8	8
4	0	1	0	21	27	0
3	0	1	25	48	13	2
2	0	4	6	4	1	0
1	3	4	11	3	1	0

TABLE 9.21
Berkeley College Applications

Department	Males		Females	
	Admitted	Refused	Admitted	Refused
A	512	313	89	19
B	353	207	17	8
C	120	205	202	391
D	138	279	131	244
E	53	138	94	299
F	22	351	24	317

One set of data collected is shown in Table 9.20. Investigate the level of agreement between the two raters by using the kappa coefficient.

9.4. Consider the following table of agreement for two raters rating a binary response.

	Yes	No	Total
Yes	15	5	20
No	5	35	40
Total	20	40	60

TABLE 9.22
Mothers' Assessments of Their Children's School Performance,
Years 1 and 2

| | Doing Well, Year 2 | | |
Doing Well, Year 1	No	Yes	Total
No	49	31	80
Yes	17	52	69
Total	66	83	149

Show that the kappa statistic for these data is identical to both their intraclass correlation coefficient and their product moment correlation coefficient.

9.5. Find the standardized residuals and the adjusted residuals for the incidence of cerebral tumors in Table 9.2. Comment on your results in the light of the nonsignificant chi-square statistic for these data.

9.6. The data in Table 9.21 classify applicants to the University of California at Berkeley by their gender, the department applied for, and whether or not they were successful in their application.

  (a) Ignoring which department is applied for, find the odds ratio for the resulting 2×2 table, gender versus application result, and also calculate its 95% confidence interval.
  (b) Calculate the chi-square test for independence of application result and department applied for, separately for men and women.
  (c) Are the results from (a) and (b) consistent and, if not, why not?

9.7 Table 9.22 shows the results of mothers' rating their children in two consecutive years, as to whether or not they were doing well at school. Carry out the appropriate test of whether there has been a change in the mothers' ratings over the 2 years.

10

Analysis of Categorical Data II: Log-Linear Models and Logistic Regression

10.1. INTRODUCTION

The two-dimensional contingency tables, which were the subject of the previous chapter, will have been familiar to most readers. Cross-classifications of more than two categorical variables, however, will not usually have been encountered in an introductory statistics course. It is such tables and their analysis that are the subject of this chapter. Two examples of *multidimensional contingency tables*, one resulting from cross-classifying three categorical variables and one from cross-classifying five such variables, appear in Tables 10.1 and 10.2. Both tables will be examined in more detail later in this chapter. To begin our account of how to deal with this type of data, we shall look at three-dimensional tables.

10.2. THREE-DIMENSIONAL CONTINGENCY TABLES

The analysis of three-dimensional contingency tables poses entirely new conceptual problems compared with the analysis of two-dimensional tables. However, the extension from tables of three dimensions to those of four or more, although

TABLE 10.1
Cross-Classification of Method of Suicide by Age and Sex

Age (Years)	Sex	Method					
		1	*2*	*3*	*4*	*5*	*6*
10–40	Male	398	121	455	155	55	124
41–70	Male	399	82	797	168	51	82
>70	Male	93	6	316	33	26	14
10–40	Female	259	15	95	14	40	38
41–70	Female	450	13	450	26	71	60
>70	Female	154	5	185	7	38	10

Note. Methods are 1, solid or liquid matter; 2, gas; 3, hanging, suffocating, or drowning; 4, guns, knives, or explosives; 5, jumping; 6, other.

TABLE 10.2
Danish Do-It-Yourself

Work	Tenure	Answer	Accommodation Type					
			Apartment			House		
			Age <30	31–45	46+	<30	31–45	46+
Skilled	Rent	Yes	18	15	6	34	10	2
		No	15	13	9	28	4	6
	Own	Yes	5	3	1	56	56	35
		No	1	1	1	12	21	8
Unskilled	Rent	Yes	17	10	15	29	3	7
		No	34	17	19	44	13	16
	Own	Yes	2	0	3	23	52	49
		No	3	2	0	9	31	51
Office	Rent	Yes	30	23	21	22	13	11
		No	25	19	40	25	16	12
	Own	Yes	8	5	1	54	191	102
		No	4	2	2	19	76	61

Note. These data arise from asking employed people whether, in the preceding year, they had carried out work on their home that they would previously have employed a craftsman to do; answers, yes or no, are cross-classified against the other four variables, which are age, accommodation type, tenure, and type of work of respondent.

TABLE 10.3
Racial Equality and the Death Penalty

Parameter	Death Penalty	Not Death Penalty
White Victim		
White defendant found guilty of murder	190	1320
Black defendant found guilty of murder	110	520
Black Victim		
White defendant found guilty of murder	0	90
Black defendant found guilty of murder	60	970
2 × 2 Table from Aggregating Data over Race of Victim		
White defendant found guilty of murder	190	1410
Black defendant found guilty of murder	170	1490

often increasing the complexity of both analysis and interpretation, presents no further new problems. Consequently, this section is concerned with only three-dimensional tables and will form the necessary basis for the discussion of models for multiway tables to be undertaken in the next section.

The first question that might be asked about a three-dimensional contingency table is, why not simply attempt its analysis by examining the two-dimensional tables resulting from summing the observed counts over one of the variables? The example shown in Table 10.3 illustrates why such a procedure is not to be recommended: it can often lead to erroneous conclusions being drawn about the data. For example, analyzing the data aggregated over the race of the victim classification gives a chi-squared statistic of 2.21 with a single degree of freedom and an associated p value of .14, implying that there is racial equality in the application of the death penalty. However, the separate analyses of the data for White victims and for Black victims lead to chi-squared values of 8.77, $p = .003$ and 5.54, $p = .019$, respectively. Claims of racial equality in the application of the death penalty now look a little more difficult to sustain.

A further example (see Table 10.4) shows that the reverse can also happen; the data aggregated over a variable can show a relationship between the remaining two variables when in fact no such relationship really exists. Here for the aggregated data the chi-square statistic is 5.26 with a single degree of freedom and an associated p value of less than .05, suggesting that infant survival *is* associated with amount of care received. For Clinic A, however, the chi-squared statistic is approximately zero as it is for Clinic B, from which the conclusion is that infant survival is *not* related to amount of care received.

TABLE 10.4
Survival of Infants and Amount of Prenatal Care

Place Where Care Received	Died		Survived	
	Less Prenatal Care	More Prenatal Care	Less Prenatal Care	More Prenatal Care
Clinic A	3	4	176	293
Clinic B	17	2	197	23

2 × 2 Table from Aggregating Data over Clinics

Amount of Prenatal Care	Infant's Survival		Total
	Died	Survived	
Less	20	373	393
More	6	316	322
Total	26	689	715

The reason for these different conclusions will become apparent later. However, these examples should make it clear why consideration of two-dimensional tables resulting from collapsing a three-dimensional table is not a sufficient procedure for analyzing the latter.

Only a single hypothesis, namely that of the independence of the two variables involved, is of interest in a two-dimensional table. However, the situation is more complex for a three-dimensional table, and several hypotheses about the three variables may have to be assessed. For example, an investigator may wish to test that some variables are independent of some others, or that a particular variable is independent of the remainder. More specifically, the following hypotheses may be of interest in a three-dimensional table.

1. There is *mutual independence* of the three variables; that is, none of the variables are related.
2. There is *partial independence*; that is, an association exists between two of the variables, both of which are independent of the third.
3. There is *conditional independence*; that is, two of the variables are independent in each level of the third, but each may be associated with the third variable (this is the situation that holds in the case of the clinic data discussed above).

In addition, the variables in a three-way contingency table may display a more complex form of association, namely what is known as a *second-order relationship*; this occurs when the degree or direction of the dependence of each pair of variables is different in some or all levels of the remaining variable. (This is analogous to the three-way interaction in a factorial design with three factors for a continuous response variable; see Chapter 4 for an example.)

As will be demonstrated later, each hypothesis is tested in a fashion exactly analogous to that used when independence is tested for in a two-dimensional table, namely by comparing the estimated expected frequencies corresponding to the particular hypothesis, with the observed frequencies, by means of the usual chi-squared statistic, X^2, or an alternative known as the *likelihood ratio statistic*, given by

$$X_L^2 = 2 \sum \text{observed} \times \ln(\text{observed/expected}), \qquad (10.1)$$

where "observed" refers to the observed frequencies in the table and "expected" refers to the estimated expected values corresponding to a particular hypothesis (see later). In many cases X^2 and X_L^2 will have similar values, but there are a number of advantages to the latter (Williams, 1976) that make it particularly suitable in the analysis of more complex contingency tables, as will be illustrated later.

Under the hypothesis of independence, the estimated expected frequencies in a two-dimensional table are found from simple calculations involving the marginal totals of frequencies as described in the previous chapter. In some cases the required expected frequencies corresponding to a particular hypothesis about the variables in a three-dimensional table can also be found from straightforward calculations on certain marginal totals (this will be illustrated in Displays 10.1 and 10.2). Unfortunately, estimated expected values in multiway tables cannot always be found so simply. For example, in a three-dimensional table, estimated expected values for the hypothesis of no second-order relationship between the three variables involves the application of a relatively complex *iterative procedure*. The details are outside the scope of this text, but they can be found in Everitt (1992). In general, of course, investigators analyzing multiway contingency tables will obtain the required expected values and associated test statistics from a suitable piece of statistical software, and so will not need to be too concerned with the details of the arithmetic.

The first hypothesis we shall consider for a three-dimensional contingency table is that the three variables forming the table are *mutually independent*. Details of how this hypothesis is formulated and tested are given in Display 10.1, and in Table 10.5 the calculation of estimated expected values, the chi-squared statistic, and the likelihood ratio statistic for the suicide data in Table 10.1 are shown. The number of degrees of freedom corresponding to each test statistic is 27 (see Everitt, 1992, for an explanation of how to determine the number of the degrees of freedom

Display 10.1

Testing the Mutual Independence Hypothesis in a Three-Dimensional Contingency Table

- Using an obvious extension of the nomenclature introduced in Chapter 9 for a two-dimensional table, we can formulate the hypothesis of mutual independence as

$$H_0 : p_{ijk} = p_{i..}p_{.j.}p_{..k},$$

where p_{ijk} represents the probability of an observation being in the ijkth cell of the table and $p_{i..}$, $p_{.j.}$, and $p_{..k}$ are the marginal probabilities of belonging to the i, j and kth categories of the three variables.

- The estimated expected values under this hypothesis when there is a sample of n observations cross-classified are

$$E_{ijk} = n\hat{p}_{i..}\hat{p}_{.j.}\hat{p}_{..k},$$

where $\hat{p}_{i..}$, $\hat{p}_{.j.}$, and $\hat{p}_{..k}$ are estimates of the corresponding probabilities.

- The intuitive (and fortunately also the maximum likelihood) estimators of the marginal probabilities are

$$\hat{p}_{i..} = n_{i..}/n, \quad \hat{p}_{.j.} = n_{.j.}/n, \quad \hat{p}_{..k} = n_{..k}/n,$$

where $n_{i..}$, $n_{.j.}$, and $n_{..k}$ are single-variable marginal totals for each variable obtained by summing the observed frequencies over the other two variables.

Display 10.2

Testing the Partial Independence Hypothesis in a Three-Dimensional Contingency Table

- With the use of the same nomenclature as before, the hypothesis of partial independence can be written in two equivalent forms as

$$H_0 : p_{ijk} = p_{..k}p_{ij.}$$

or

$$H_0 : p_{i.k} = p_{i..}p_{..k} \quad \text{and} \quad p_{.jk} = p_{.j.}p_{..k}.$$

- The estimators of the probabilities involved are

$$\hat{p}_{..k} = n_{..k}/n, \qquad \hat{p}_{ij.} = n_{ij.}/n,$$

where $n_{ij.}$ represents the two-variable marginal totals obtained by summing the observed frequencies over the third variable.

- Using these probability estimates leads to the following estimated expected values under this hypothesis:

$$E_{ijk} = n_{..k}n_{ij.}/n.$$

---

corresponding to a particular hypothesis). Clearly, the three variables used to form Table 10.1 are not mutually independent.

Now consider a more complex *partial independence hypothesis* about the three variables in a three-dimensional contingency table, namely that variable 1 (say) is independent of variable 2, and that variables 2 and 3 are also unrelated. However, an association between variable 1 and 3 is allowed. Display 10.2 gives details of

TABLE 10.5
Testing Mutual Independence for the Suicide Data

- For the suicide data, the estimated expected value, E_{111}, under the hypothesis of mutual independence is obtained as

$$E_{111} = 5305 \times 3375/5305 \times 1769/5305 \times 1735/5305 = 371.89.$$

- Other estimated expected values can be found in a similar fashion, and the full set of values for the suicide data under mutual independence are as follows.

Age	Sex	1	2	Method 3	4	5	6
10–40	Male	371.9	51.3	487.5	85.5	59.6	69.6
41–70	Male	556.9	76.9	730.0	128.0	89.3	104.2
>70	Male	186.5	25.7	244.4	42.9	29.9	34.9
10–40	Female	212.7	29.4	278.8	48.9	34.1	39.8
41–70	Female	318.5	44.0	417.5	73.2	51.0	59.6
>70	Female	106.6	14.7	139.8	24.5	17.2	20.0

- The values of the two possible test statistics are

$$X^2 = 747.37, \quad X_L^2 = 790.30.$$

- The mutual independence hypothesis has 27 degrees of freedom.
- Note that the single-variable marginal totals of the estimated expected values under the hypothesis of mutual independence are equal to the corresponding marginal totals of the observed values, for example,

$$n_{1..} = 398 + 399 + 93 + 259 + 450 + 154 = 1753,$$
$$E_{1..} = 371.9 + 556.9 + 186.5 + 212.7 + 318.5 + 106.6 = 1753.$$

how this hypothesis is formulated and the calculation of the estimated expected values. In Table 10.6 the partial independence hypothesis is tested for the suicide data. Clearly, even this more complicated hypothesis is not adequate for these data. A comparison of the observed values with the estimates of the values to be expected under this partial independence hypothesis shows that women in all age groups are underrepresented in the use of guns, knives, or explosives (explosives!) to perform the tragic task. (A more detailed account of how best to compare observed and expected values is given later.)

Finally, consider the hypothesis of no second-order relationship between the variables in the suicide data. This allows each pair of variables to be associated, but it constrains the degree and direction of the association to be the same in each level of the third variable. Details of testing the hypothesis for the suicide data are given in Table 10.7. Note that, in this case, estimated expected values cannot be found directly from any set of marginal totals. They are found from the iterative procedure referred to earlier. (The technical reason for requiring an iterative process is that, in this case, the maximum likelihood equations from

TABLE 10.6

Testing the Partial Independence Hypothesis for the Suicide Data

---

- Here we wish to test whether method of suicide is independent of sex, and that age and sex are also unrelated. However, we wish to allow an association between age and method.
- The estimated expected value E_{111} under this hypothesis is found as

$$E_{111} = (3375 \times 657)/(5305) = 417.98.$$

- Other estimated expected values can be found in a similar fashion, and the full set of such values under the partial independence hypothesis is as follows.

				Methods			
Age	Sex	1	2	3	4	5	6
10–40	Male	418.0	86.5	349.9	107.5	60.44	103.1
41–70	Male	540.1	60.4	793.3	123.4	77.6	90.3
>70	Male	157.1	7.0	318.7	25.5	40.7	15.3
10–40	Female	239.0	49.5	200.1	61.5	34.6	59.0
41–70	Female	308.9	34.6	453.7	70.6	44.4	51.7
>70	Female	89.9	4.0	182.3	14.6	23.3	8.7

- Note that in this case the marginal totals, $n_{ij.}$ and $E_{ij.}$, are equal; for example, $n_{11.} = 398 + 259 = 657$ and $E_{11.} = 418.0 + 239.0 = 657$.
- The values of the two test statistics are

$$X^2 = 485.3, \quad X_L^2 = 520.4.$$

- These statistics have 17 degrees of freedom under the partial independence hypothesis.

---

which estimates arise have no explicit solution. The equations have to be solved iteratively.) Both test statistics are nonsignificant, demonstrating that this particular hypothesis is acceptable for the suicide data. Further comments on this result will be given in the next section.

10.3. MODELS FOR CONTINGENCY TABLES

Statisticians are very fond of models! In the previous chapters the majority of analyses have been based on the assumption of a suitable model for the data of interest. The analysis of categorical data arranged in the form of a multiway frequency table may also be based on a particular type of model, not dissimilar to those used in the analysis of variance. As will be seen, each particular model corresponds to a specific hypothesis about the variables forming the table, but the advantages to be gained from a model-fitting procedure are that it provides a systematic approach to the analysis of complex multidimensional tables and, in addition, gives estimates of the magnitude of effects of interest. The models

TABLE 10.7
Testing the Hypothesis of No Second-Order Relationship Between the Variables in the Suicide Data

- The hypothesis of interest is that the association between any two of the variables does not differ in either degree or direction in each level of the remaining variable.
- More specifically, this means that the odds ratios corresponding to the 2 × 2 tables that arise from the cross-classification of pairs of categories of two of the variables are the same in all levels of the remaining variable. (Details are given in Everitt, 1992.)
- Estmates of expected values under this hypothesis cannot be found from simple calculations on marginal totals of observed frequencies as in Displays 10.1 and 10.2. Instead, the required estimates have to be obtained iteratively by using a procedure described in Everitt (1992).
- The estimated expected values derived form this iterative procedure are as follows.

		Method					
Age	Sex	1	2	3	4	5	6
10–40	Male	410.9	122.7	439.2	156.4	56.8	122.0
41–70	Male	379.4	77.6	819.9	166.3	51.1	84.7
>70	Male	99.7	8.7	308.9	33.4	24.1	13.3
10–40	Female	246.1	13.3	110.8	12.6	38.2	40.0
41–70	Female	496.6	17.4	427.1	27.7	70.9	57.3
>70	Female	147.3	2.3	192.1	6.6	39.9	10.7

- The two test statistics take the values

$$X^2 = 15.40, \quad X_L^2 = 14.90.$$

- The degrees of freedom for this hypothesis are 10.

used for contingency tables can be introduced most simply (if a little clumsily) in terms of a two-dimensional table; details are given in Display 10.3. The model introduced there is analogous to the model used in a two-way analysis of variance (see Chapter 4), but it differs in a number of aspects.

1. The data now consist of counts, rather than a score for each subject on some dependent variable.
2. The model does not distinguish between independent and dependent variables. All variables are treated alike as response variables whose mutual associations are to be explored.
3. Whereas a linear combination of parameters is used in the analysis of variance and regression models of previous chapters, in multiway tables the natural model is *multiplicative* and hence logarithms are used to obtain a model in which parameters are combined additively.
4. In previous chapters the underlying distribution assumed for the data was the normal; with frequency data the appropriate distribution is *binomial* or *multinomial* (see the glossary in Appendix A).

Display 10.3
Log-Linear Model for a Two-Dimensional Contingency Table with r Rows and c Columns

- Again the general model considered in previous chapters, that is,

$$\text{observed response} = \text{expected response} + \text{error},$$

is the starting point.
- Here the observed response is the observed count, n_{ij} in a cell of the table, and the expected response is the frequency to be expected under a particular hypothesis, F_{ij}. Hence

$$n_{ij} = F_{ij} + \text{error}.$$

- Unlike the corresponding terms in models discussed in previous chapters, the error terms here will not be normally distributed. Appropriate distributions are the binomial and multinomial (see glossary in Appendix A).
- Under the independence hypothesis, the population frequencies, F_{ij}, are given by

$$F_{ij} = n p_{i.} p_{.j},$$

which can be rewritten, using an obvious dot notation, as

$$F_{ij} = n \times \frac{F_{i.}}{n} \times \frac{F_{.j}}{n} = \frac{F_{i.} F_{.j}}{n}.$$

- When logarithms are taken, the following linear model for the expected frequencies is arrived at:

$$\ln F_{ij} = \ln F_{i.} + \ln F_{.j} - \ln n.$$

- By some simple algebra (it really *is* simple, but see Everitt, 1992, for details), the model can be rewritten in the form

$$\ln F_{ij} = u + u_{1(i)} + u_{2(j)},$$

where

$$u = \frac{\sum_{i=1}^{r} \sum_{j=1}^{c} \ln F_{ij}}{rc},$$

$$u_{1(i)} = \frac{\sum_{j=1}^{c} \ln F_{ij}}{c} - u,$$

$$u_{2(j)} = \frac{\sum_{i=1}^{r} \ln F_{ij}}{r} - u.$$

- The form of the model is now very similar to those used in the analysis of variance (see Chapters 3 and 4). Consequently, ANOVA terms are used for the parameters, and u is said to represent an overall mean effect, $u_{1(i)}$ is the main effect of category i of the row variable, and $u_{2(j)}$ is the main effect of the jth category of the columns variable.
- The main effect parameters are defined as deviations of row or columns means of log frequencies from the overall mean. Therefore, again using an obvious dot notation,

$$u_{1(.)} = 0, \quad u_{2(.)} = 0.$$

- The values taken by the main effects parameters in this model simply reflect differences between the row or columns marginal totals and so are of little concern in the context of the analysis of contingency tables. They could be estimated by replacing the F_{ij} in the formulas above with the estimated expected values, E_{ij}.

- The log-linear model can be fitted by estimating the parameters, and hence the expected frequencies, and comparing these with the observed values, using either the chi-squared or likelihood ratio test statistics. This would be exactly equivalent to the usual procedure for testing independence in a two-dimensional contingency table as described in the previous chapter.
- If the independence model fails to give a satisfactory fit to a two-dimensional table, extra terms must be added to the model to represent the association between the two variables. This leads to a further model,

$$\ln F_{ij} = u + u_{1(i)} + u_{2(j)} + u_{12(ij)},$$

where the parameters $u_{12(ij)}$ model the association between the two variables.

- This is known as the *saturated model* for a two-dimensional contingency table because the number of parameters in the model is equal to the number of independent cells in the table (see Everitt, 1992, for details). Estimated expected values under this model would simply be the observed frequencies themselves, and the model provides a perfect fit to the observed data, but, of course, no simplification in description of the data.
- The interaction parameters $u_{12(ij)}$ are related to odds ratios (see Exercise 10.7).

Now consider how the log-linear model in Display 10.3 has to be extended to be suitable for a three-way table. The saturated model will now have to contain main effect parameters for each variable, parameters to represent the possible associations between each pair of variables, and finally parameters to represent the possible second-order relationship between the three variables. The model is

$$\ln F_{ijk} = u + u_{1(i)} + u_{2(j)} + u_{3(k)} + u_{12(ij)}$$
$$+ u_{13(ik)} + u_{23(jk)} + u_{123(ijk)}. \qquad (10.2)$$

The parameters in this model are as follows.

1. u is the overall mean effect.
2. $u_{1(i)}$ is the main effect of variable 1.
3. $u_{2(j)}$ is the main effect of variable 2.
4. $u_{3(k)}$ is the main effect of variable 3.
5. $u_{12(ij)}$ is the interaction between variables 1 and 2.
6. $u_{13(ik)}$ is the interaction between variables 1 and 3.
7. $u_{23(jk)}$ is the interaction between variables 2 and 3.
8. $u_{123(ijk)}$ is the second-order relationship between the three variables.

The purpose of modeling a three-way table would be to find the *unsaturated* model with fewest parameters that adequately predicts the observed frequencies. As a way to assess whether some simpler model would fit a given table, partic- ular parameters in the saturated model are set to zero and the reduced model is assessed for fit. However, it is important to note that, in general, attention must be restricted to what are known as *hierarchical models*. These are such that whenever a higher-order effect is included in a model, the lower-order effects composed from variables in the higher effect are also included. For example, if terms u_{123} are included, so also must terms u_{12}, u_{13}, u_1, u_2, and u_3. Therefore, models such as

$$\ln F_{ijk} = u + u_{2(j)} + u_{3(k)} + u_{123(ijk)} \tag{10.3}$$

are not permissible. (This restriction to hierarchical models arises from the constraints imposed by the maximum likelihood estimation procedures used in fitting log-linear models, details of which are too technical to be included in this text. In practice, the restriction is of little consequence because most tables *can* be described by a series of hierarchical models.)

Each model that can be derived from the saturated model for a three-dimensional table is equivalent to a particular hypothesis about the variables forming the table; the equivalence is illustrated in Display 10.4. Particular points to note about the material in this display are as follows.

1. The first three models are of no consequence in the analysis of a three-dimensional table. Model 4 is known as the *minimal* model for such a table.

2. The *fitted marginals* or bracket notation is frequently used to specify the series of models fitted when a multidimensional contingency table is examined. The notation reflects the fact noted earlier that, when testing particular hypotheses about a multiway table (or fitting particular models), certain marginal totals of the estimated expected values are constrained to be equal to the corresponding marginals of the observed values. (This arises because of the form of the maximum likelihood equations.) The terms used to specify the model with the bracket notation are the marginals fixed by the model.

To illustrate the use of log-linear models in practice, a series of such models will be fitted to the suicide data given in Table 10.1. Details are given in Table 10.8. The aim of the procedure is to arrive at a model that gives an adequate fit to the observed data and, as shown in Table 10.8, differences in the likelihood ratio statistic for different models are used to assess whether models of increasing complexity (larger number of parameters) are needed.

The results given in Table 10.8 demonstrate that only model 7 provides an adequate fit for the suicide data. This model states that the association between age and method of suicide is the same for males and females, and that the association between sex and method is the same for all age groups. The parameter estimates (and the ratio of the estimates to their standard errors) for the fitted model are given in Table 10.9. The estimated main effects parameters are not of great interest; their values simply reflect differences between the marginal totals of the categories of each variable. For example, the largest effect for method, $\hat{u}_{1(2)} = 1.55$, arises only

Display 10.4

Hierarchical Models for a General Three-Dimensional Contingency Table

---

- A series of possible log-linear models for a three-way contingency table is as follows.

Log-Linear Model	Bracket Notation
1. $\ln F_{ijk} = u$	
2. $\ln F_{ijk} = u + u_{1(i)}$	[1]
3. $\ln F_{ijk} = u + u_{1(i)} + u_{2(j)}$	[1],[2]
4. $\ln F_{ijk} = u + u_{1(i)} + u_{2(j)} + u_{3(k)}$	[1],[2],[3]
5. $\ln F_{ijk} = u + u_{1(i)} + u_{2(j)} + u_{3(k)} + u_{12(ij)}$	[12],[3]
6. $\ln F_{ijk} = u + u_{1(i)} + u_{2(j)} + u_{3(k)} + u_{12(ij)} + u_{13(ik)}$	[12], [13]
7. $\ln F_{ijk} = u + u_{1(i)} + u_{2(j)} + u_{3(k)} + u_{12(ij)} + u_{13(ik)}$ $+ u_{23(jk)}$	[12], [13], [23]
8. $\ln F_{ijk} = u + u_{1(i)} + u_{2(j)} + u_{3(k)} + u_{12(ij)} + u_{13(ik)}$ $+ u_{23(jk)} + u_{123(ijk)}$	[123]

- The hypotheses corresponding to the first seven models are as follows.
  1. All frequencies are the same.
  2. Marginal totals for variable 2 and variable 3 are equal.
  3. Marginal totals for variable 3 are equal. (Because these first three models do not allow the observed frequencies to reflect observed differences in the marginal totals of each variable, they are of no real interest in the analysis of three-dimensional contingency tables.)
  4. The variables are mutually independent.
  5. Variables 1 and 2 are associated and both are independent of variable 3.
  6. Variables 2 and 3 are conditionally independent given variable 1.
  7. There is no second-order relationship between the three variables.
  8. Model 8 is the saturated model for a three-dimensional table.

---

because more people use hanging, suffocating, or drowning as a method of suicide than the other five possibilities. The estimated interaction parameters are of more interest, particularly those for age and method and for sex and method. For example, the latter reflect that males use "solid" and "jump" less and women use them more than if sex was independent of method. The reverse is true for "gas" and "gun".

As always when models are fitted to observed data, it is essential to examine the fit in more detail than is provided by a single goodness-of-fit statistic such as the likelihood ratio criterion. With log-linear models, differences between the observed (O) and estimated expected (E) frequencies form the basis for this more detailed examination, generally using the standardized residual calculated as

$$\text{standardized residual} = (O - E)/\sqrt{E}. \qquad (10.4)$$

The residuals for the final model selected for the suicide data are given in Table 10.10. All of the residuals are small, suggesting that the chosen model does give an adequate representation of the observed frequencies. (A far fuller account of log-linear models is given in Agresti, 1996.)

TABLE 10.8
Log-Linear Models for Suicide Data

- The goodness of fit of a series of log-linear models is given below (variable 1 is method, variable 2 is age, and variable 3 is sex).

Model	DF	X_L^2	p
[1],[2],[3]	27	790.3	<.001
[12], [3]	17	520.4	<.001
[13], [2]	22	424.6	<.001
[23], [1]	25	658.9	<.001
[12], [13]	12	154.7	<.001
[12], [23]	15	389.0	<.001
[12], [13], [23]	10	14.9	.14

- The difference in the likelihood ratio statistic for two possible models can be used to choose between them. For example, the mutual independence model, [1], [2], [3], and the model that allows an association between variables 1 and 2, [12], [3], have $X_L^2 = 790.3$ with 27 degrees of freedom and $X_L^2 = 520.4$ with 17 degrees of freedom, respectively. The hypothesis that the extra parameters in the more complex model are zero, that is, $H_0 : u_{12(ij)} = 0$ for all i and j ($u_{12} = 0$ for short), is tested by the difference in the two likelihood ratio statistics, with degrees of freedom equal to the difference in the degrees of freedom of the two models.
- Here this leads to a value of 269.9 with 10 degrees of freedom. The result is highly significant and the second model provides a significantly improved fit compared with the mutual independence model.
- A useful way of judging a series of log-linear models is by means of an analog of the square of the multiple correlation coefficient as used in multiple regression (see Chapter 6). The measure L is defined as

$$L = \frac{X_L^2(\text{baseline model}) - X_L^2(\text{model of interest})}{X_L^2(\text{baseline model})}.$$

- L lies in the range (0,1) and indicates the percentage improvement in goodness of fit of the model being tested over the baseline model.
- The choice of baseline model is not fixed. It will often be the mutual independence model, but it could be the simpler of two competing models.
- For comparing the mutual independence and no second-order relationship models on the suicide data, L, is $L = (790.3 - 14.9)/790.3 = 98.1\%$.

10.4. LOGISTIC REGRESSION
FOR A BINARY RESPONSE VARIABLE

In many multidimensional contingency tables (I am tempted to say in almost all), there is one variable that can properly be considered a response, and it is the relationship of this variable to the remainder, which is of most interest. Situations in which the response variable has two categories are most common, and it is these that are considered in this section. When these situations are introduced

TABLE 10.9
Parameter Estimates in the Final Model Selected for the Suicide Data

- The final model selected for the suicide data is that of no second-order relationship between the three variables, that is, model 7 in Display 10.4.
- The estimated main effect parameters are as follows.

Variable	Parameter	Category	Estimate	Estimate/SE
Method	$u_{1(1)}$	Solid	1.33	34.45
	$u_{1(2)}$	Gas	−1.27	−11.75
	$u_{1(3)}$	Hang	1.55	41.90
	$u_{1(4)}$	Gun	−0.64	−8.28
	$u_{1(5)}$	Jump	−0.42	−7.00
	$u_{1(6)}$	Other	−0.55	−7.64
Age	$u_{2(1)}$	10–40	0.25	7.23
	$u_{2(2)}$	41–70	0.56	16.77
	$u_{2(3)}$	>70	−0.81	−16.23
Sex	$u_{3(1)}$	Male	0.41	15.33
	$u_{3(2)}$	Female	−0.41	−15.33

(Note that the parameter estimates for each variable sum to zero.)
- The estimated interaction parameters for method and age and their ratio to the corresponding standard error (in parentheses) are as follows.

			Method			
Age	Solid	Gas	Hanging	Gun	Jump	Other
10–40	−0.0 (−0.5)	0.5 (4.9)	−0.6 (−14.2)	−0.0 (−0.4)	−0.2 (−2.5)	0.3 (4.2)
41–70	−0.1 (−1.09)	0.1 (1.0)	0.1 (1.7)	0.1 (1.3)	−0.3 (−3.4)	0.0 (0.4)
>70	0.1 (1.1)	−0.6 (−3.6)	0.5 (9.6)	−0.1 (−0.6)	0.5 (4.8)	−0.4 (−3.0)

- The estimated interaction parameters for method and sex and the ratios of these estimates to their standard errors (in parentheses) are as follows.

			Method			
Sex	Solid	Gas	Hanging	Gun	Jump	Other
Male	−0.4 (−13.1)	0.4 (5.3)	0.0 (0.3)	0.6 (8.4)	−0.5 (−8.6)	−0.1 (−2.2)
Female	0.4 (13.1)	−0.4 (−5.3)	−0.0 (0.3)	−0.6 (−8.4)	0.5 (8.6)	0.1 (2.2)

- The estimated interaction parameters for age and sex and the ratios of these estimates to their standard errors (in parentheses) are as follows.

		Age	
Sex	10–40	41–70	>70
Male	0.3 (11.5)	−0.1 (−4.5)	−0.2 (−6.8)
Female	−0.3 (−11.5)	0.1 (4.5)	0.2 (6.8)

TABLE 10.10
Standardized Residuals from the Final Model Fitted to Suicide Data

		Method					
Age	Sex	1	2	3	4	5	6
10–40	Male	−0.6	−0.2	0.8	−0.1	−0.2	0.2
41–70	Male	1.0	0.5	−0.8	0.1	−0.0	−0.3
>70	Male	−0.7	−0.9	0.4	−0.1	0.4	0.2
10–40	Female	0.8	0.5	−1.5	0.4	0.3	−0.3
41–70	Female	−0.9	−1.1	1.1	−0.3	0.0	0.4
>70	Female	0.6	1.8	−0.5	0.1	−0.3	−0.2

in the context of a multidimensional contingency table, it might be thought that our interest in them is confined to only categorical explanatory variables. As we shall see in the examples to come, this is *not* the case, and explanatory variables might be a mixture of categorical and continuous. In general the data will consist of either obervations from individual subjects having the value of a zero–one response variable and associated explanatory variable values, or these observations are *grouped* contingency table fashion, with counts of the number of zero and one values of the response in each cell.

Modeling the relationship between a response variable and a number of explanatory variables has already been considered in some detail in Chapter 6 and in a number of other chapters, so why not simply refer to the methods described previously? The reason lies with the nature of the response variable. Explanations will become more transparent if considered in the context of an example, and here we will use the data shown in Table 10.11. These data arise from a study of a psychiatric screening questionnaire called the General Health Questionnaire (GHQ) (see Goldberg, 1972).

How might these data be modeled if interest centers on how "caseness" is related to gender and GHQ score? One possibility that springs to mind would be to consider modeling the probability, p, of being a case. In terms of the general model encountered in earlier chapters and in Display 10.1, that is,

$$\text{observed response} = \text{expected response} + \text{error,} \qquad (10.5)$$

this probability would be the expected response and the corresponding observed response would be the proportion of individuals in the sample categorized as cases.

TABLE 10.11
GHQ Data

GHQ Score	Sex	No. of Cases	No. of Noncases
0	F	4	80
1	F	4	29
2	F	8	15
3	F	6	3
4	F	4	2
5	F	6	1
6	F	3	1
7	F	2	0
8	F	3	0
9	F	2	0
10	F	1	0
0	M	1	36
1	M	2	25
2	M	2	8
3	M	1	4
4	M	3	1
5	M	3	1
6	M	2	1
7	M	4	2
8	M	3	1
9	M	2	0
10	M	2	0

Note. F, female; M, male.

Therefore, a possible model is

$$p = \beta_0 + \beta_1 \text{sex} + \beta_2 \text{GHQ} \tag{10.6}$$

and

$$\text{observed response} = p + \text{error.} \tag{10.7}$$

To simplify things for the moment, let's ignore gender and fit the model

$$p = \beta_0 + \beta_1 \text{GHQ} \tag{10.8}$$

by using a least-squares approach as described in Chapter 6. Estimates of the two parameters, estimated standard errors, and predicted values of the response

TABLE 10.12

Linear Regression Model for GHQ Data with GHQ Score as the Single
Explanatory Variable

Parameter	Estimate	SE	t
β_0(intercept)	0.136	0.065	2.085
β_1(GHQ)	0.096	0.011	8.698

Predicted Values of p from This Model

Observation No.	Predicted Prob.	Observed Prob.
1	0.136	0.041
2	0.233	0.100
3	0.328	0.303
4	0.425	0.500
5	0.521	0.700
6	0.617	0.818
7	0.713	0.714
8	0.809	0.750
9	0.905	0.857
10	1.001	1.000
11	1.097	1.000

are shown in Table 10.12. Immediately a problem becomes apparent—two of the
predicted values are greater than one, but the response is a probability constrained
to be in the interval (0,1). Thus using a linear regression approach here can lead
to fitted probability values outside the range 0,1. An additional problem is that the
error term in the linear regression model is assumed to have a normal distribution;
this is clearly not suitable for a binary response.

It is clearly not sensible to contemplate using a model, which is known *a priori*
to have serious disadvantages, and so we need to consider an alternative approach
to linear regression for binary responses. That most frequently adopted is the
linear logistic model, or *logistic model* for short. Now a *transformed* value of p is
modeled rather than p directly, and the transformation chosen ensures that fitted
values of p lie in the interval (0,1). Details of the logistic regression model are
given in Display 10.5.

Fitting the logistic regression model to the GHQ data, disregarding gender,
gives the results shown in Table 10.13. Note that now all the predicted values are
satisfactory and lie between 0 and 1. A graphical comparison of the fitted linear
and logistic regressions is shown in Figure 10.1. We see that in addition to the
problems noted earlier with using the linear regression approach here, this model
provides a very poor description of the data.

Display 10.5
The Logistic Regression Model

- The *logistic transformation*, λ, of a probability, p, is defined as follows:

$$\lambda = \ln[p/(1 - p)].$$

  In other words, λ is the logarithm of the odds ratio for the response variable.
- As p varies from 0 to 1, λ varies between $-\infty$ and ∞.
- The logistic regression model is a linear model for λ, that is,

$$\lambda = \ln[p/(1 - p)] = \beta_0 + \beta_1 x_1 + \beta_2 x_2 + \cdots + \beta_q x_q,$$

  where $x_1, x_2, \ldots, x_q$ are the q explanatory variables of interest.
- Modeling the logistic transformation of p rather than p itself avoids possible problems of finding fitted values outside their permitted range. ($\ln[p/(1 - p)]$ is often written as logit(p) for short.)
- The parameters in the model are estimated by maximum likelihood (see Collett, 1991, for details).
- The parameters in the model can be interpreted as the change in the ln(odds) of the response variable produced by a change of one unit in the corresponding explanatory variable, conditional on the other variables' remaining constant.
- It is sometimes convenient to consider the model as it represents p itself, that is,

$$p = \frac{\exp(\beta_0 + \beta_1 x_1 + \cdots + \beta_q x_q)}{1 + \exp(\beta_0 + \beta_1 x_1 + \cdots + \beta_q x_q)}.$$

- There are a number of summary statistics that measure the discrepancy between the observed proportions of "success" (i.e., the "one" category of the response variable, say), and the fitted proportions from the logistic model for the true success probability p. The most common is known as the *deviance*, D, and it is given by

$$D = 2 \sum_{i=1}^{k} \left[y_i \ln \left(\frac{y_i}{\hat{y}_i} \right) + (n_i - y_i) \ln \left(\frac{n_i - y_i}{n_i - \hat{y}_i} \right) \right],$$

  where y_i is the number of success in the ith category of the observations, n_i is the total number of responses in the ith category, and $\hat{y}_i = n_i \hat{p}_i$ where $\hat{p}_i$ is the predicted success probability for this category. (We are assuming here that the raw data have been collected into categories as in the GHQ data and the Danish do-it-yourself data in the text. When this is not so and the data consist of the original zeros and ones for the response, then the deviance *cannot* be used as a measure of fit; see Collett, 1991, for an explanation of why not.)
- The deviance is distributed as chi-squared and differences in deviance values can be used to assess competing models; see the examples in the text.

The estimated regression coefficient for GHQ score in the logistic regression is 0.74. Thus the log (odds) of being a case increases by 0.74 for a unit increase in GHQ score. An approximate 95% confidence interval for the regression coefficient is

$$0.74 \pm 1.96 \times 0.09 = (0.56, 0.92). \qquad (10.9)$$

TABLE 10.13

Logistic Regression Model for GHQ Data with GHQ Score
as the Single Explanatory Variable

Parameter	Estimate	SE	t
β_0 (intercept)	−2.711	0.272	−9.950
β_1 (GHQ)	0.736	0.095	7.783

Observation No.	Predicted Prob.	Observed Prob.
1	0.062	0.041
2	0.122	0.100
3	0.224	0.303
4	0.377	0.500
5	0.558	0.700
6	0.725	0.818
7	0.846	0.714
8	0.920	0.750
9	0.960	0.857
10	0.980	1.000
11	0.991	1.000

However, such results given as they are in terms of log (odds) are not immediately helpful. Things become better if we translate everything back to odds, by
exponentiating the various terms. So, $\exp(0.74) = 2.08$ represents the increase in
the odds of being a case when the GHQ score increases by one. The corresponding
confidence interval is $[\exp(0.56), \exp(0.092)] = [1.75, 2.51]$.

Now let us consider a further simple model for the GHQ data, namely a logistic
regression for caseness with only gender as an explanatory variable. The results
of fitting such a model are shown in Table 10.14. Again, the estimated regression
coefficient for gender, −0.04, represents the change in log (odds) (here a decrease)
as the explanatory variable increases by one. For the dummy variable coding sex,
such a change implies that the observation arises from a man rather than a woman.
Transferring back to odds, we have a value of 0.96 and a 95% confidence interval
of (0.55, 1.70). Because this interval contains the value one, which would indicate
the independence of gender and caseness (see Chapter 9), it appears that gender
does not predict caseness for these data.

If we now look at the 2 × 2 table of caseness and gender for the GHQ data,
that is,

	Sex	
	Female	Male
Case	43	25
Not case	131	79

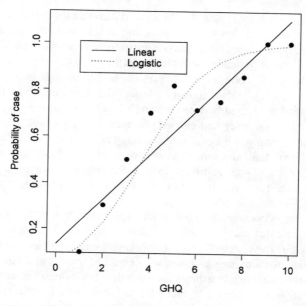

FIG. 10.1. Fitted linear and logistic regression models for the probability of being a case as a function of the GHQ score for the data in Table 10.11

TABLE 10.14
Logistic Regression Model for GHQ Data with Gender as
the Single Explanatory Variable

Parameter	Estimate	SE	t
β_0 (intercept)	−1.114	0.176	−6.338
β_0 (sex)	−0.037	0.289	−0.127

we see that the odds ratio is $(131 \times 25)/(43 \times 79) = 0.96$, the same result as given by the logistic regression. For a single binary explanatory variable, the estimated regression coefficient is simply the log of the odds ratio from the 2×2 table relating the response variable to the explanatory variable. (Readers are encouraged to confirm that the confidence interval found from the logistic regression is the same as would be given by using the formula for the variance of the log of the odds ratio given in the previous chapter. As a further exercise, readers might fit a logistic regression model to the GHQ data, which includes both GHQ score and gender as explanatory variables. If they do, they will notice that the estimated regression coefficient for sex is no longer simply the log of the odds ratio from the 2×2 table above, because of the conditional nature of these regression coefficients when other variables are included in the fitted model.)

As a further more complex illustration of the use of logistic regression, the method is applied to the data shown in Table 10.2. The data come from a sample of employed men aged between 18 and 67 years; who were asked whether, in the preceding year, they had carried out work in their home that they would have previously employed a craftsman to do. The response variable here is the answer (yes/no) to that question. There are four categorical explanatory variables: (1) *age*: under 30, 31–45, and over 45; (2) *accommodation type:* apartment or house; (3) *tenure*: rent or own; and (4) *work of respondent*: skilled, unskilled, or office.

To begin, let us consider only the single explanatory variable, work of respondent, which is a categorical variable with three categories. In Chapter 6, when the use of multiple regression with categorical explanatory variables was discussed, it was mentioned that although such variables could be used, some care was needed in how to deal with them when they have more than two categories, as here. Simply coding the categories as, say, 1, 2, and 3 for the work variable would really not do, because it would imply equal intervals along the scale. The proper way to handle a nominal variable with more that $k > 2$ categories would be to recode it as a number of dummy variables. Thus we will recode work in terms of two dummy variables, work1 and work2, defined as follows.

Work	Work1	Work2
Skilled	0	0
Unskilled	1	0
Office	0	1

The results of the logistic regression model of the form $\text{logit}(p) = \beta_0 + \beta_1 \text{work1} + \beta_2 \text{work2}$ are given in Table 10.15. A cross-classification of work

TABLE 10.15
Logistic Regression for Danish Do-It-Yourself Data with
Work as the Single Explanatory Variable

Parameter	Estimate	SE	t
β_0 (intercept)	0.706	0.112	6.300
β_1 (work1)	−0.835	0.147	−5.695
β_2 (work2)	−0.237	0.134	−1.768

TABLE 10.16
Cross-Classification of Work Against Response for Danish
Do-It-Yourself Data

		Work		
Response	Office	Skilled	Unskilled	Total
No	301	119	239	659
Yes	481	241	210	932
Total	782	360	449	1591

against the response is shown in Table 10.16. From this table we can extract the following pair of 2×2 tables:

Response	Skilled	Unskilled
No	119	239
Yes	241	210

odds ratio $= (119 \times 210)/(241 \times 239) = 0.434$, log (odds ratio) $= -0.835$.

Response	Skilled	Office
No	119	301
Yes	241	481

odds ratio $= (119 \times 481)/(241 \times 301) = 0.789$, log (odds ratio) $= -0.237$.

We see that the coding used produces estimates for the regression coefficients of work1 and work2 that are equal to the log (odds ratio) from comparing skilled and unskilled and skilled and office. (Readers are encouraged to repeat this exercise; use the age variable represented also as two dummy variables.)

The results from fitting the logistic regression model with all four explanatory variables to the Danish do-it-yourself data are shown in Table 10.17. Work has been recoded as work1 and work2 as shown above, and age has been similarly coded in terms of two dummy variables, age1 and age2. So the model fitted is $logit(p) = \beta_0 + \beta_1 work1 + \beta_2 work2 + \beta_3 age1 + \beta_4 age2 + \beta_5 tenure + \beta_6 type$. Note that the coefficients for work1 and work2 are similar, but not identical to those in Table 10.15. They can, however, still be interpreted as log (odds ratios), but after the effects of the other three explanatory variables are taken into account. The results in Table 10.17 appear to imply that work, age, and tenure are the three most important explanatory variables for predicting the probability of answering yes to the question posed in the survey. However, the same caveats apply to logistic regression as were issued in the case of multiple linear regression in

TABLE 10.17

Results from Fitting the Logistic Model to Danish Do-It-Yourself
Data by Using All Four Observed Explanatory Variables

Parameter	Estimate	SE	t
β_0(intercept)	0.306	0.154	1.984
β_1(work1)	−0.763	0.152	−5.019
β_2(work2)	−0.305	0.141	−2.168
β_3(age1)	−0.113	0.137	−0.825
β_4(age2)	−0.437	0.141	−3.106
β_5(tenure)	1.016	0.138	7.368
β_6(type)	−0.002	0.147	−0.017

Chapter 6—these regression coefficients and the associated standard errors are
estimated, *conditional* on the other variables' being in the model. Consequently,
the t values give only a rough guide to which variables should be included in a
final model.

Important subsets of explanatory variables in logistic regression are often selected by using a similar approach to the forward, backward, and stepwise procedures described in Chapter 6, although the criterion for deciding whether or not
a candidate variable should be added to, or excluded from, an existing model is
different, now usually involving the deviance index of goodness of fit described
in Display 10.5. Many statistical packages have automatic variable selection procedures to be used with logistic regression, but here we shall try to find a suitable
model for the Danish do-it-youself data in a relatively informal manner, by examining deviance differences as variables are added to a current model. Details are
given in Table 10.18.

It is clear from the results in Table 10.18 that Tenure, Work, and Age are all
required in a final model. The parameter estimates, and so on, for this model are
shown in Table 10.19, along with the observed and predicted probabilities of giving
a yes answer to the question posed in the survey. Explicitly, the final model is

$$\ln \frac{p}{1-p} = 0.30 + 1.01\text{tenure} - 0.76\text{work1} - 0.31\text{work2}$$

$$- 0.11\text{age1} - 0.43\text{age2}. \tag{10.10}$$

Before trying to interpret this model, it would be wise to look at some diagnostics
that will indicate any problems it has, in a similar way to that described in Chapter 6
for multiple linear regression. Two useful diagnostics for logistic regression are described in Display 10.6. Helpful plots of these two diagnostics are shown in Figure
10.2. Figure 10.2(a) shows the deviance residuals plotted against the fitted values,

TABLE 10.18
Comparing Logistic Models for the Danish Do-It-Yourself Data

- We will start with a model including only tenure as an explanatory variable and then add explanatory variables in an order suggested by the t statistics from Table 10.17.
- Differences in the deviance values for the various models can be used to assess the effect of adding the new variable to the existing model. These differences can be tested as chi-squares with degrees of freedom equal to the difference in the degrees of freedom of the two models that are being compared.
- We can arrange the calculations in what is sometimes known as an *analysis of deviance table*

Model	Deviance	DF	Deviance diff.	DF diff.	p
Tenure	72.59	34			
Tenure + Work	40.61	32	31.98	2	<.0001
Tenure + Work + Age	29.67	30	10.94	2	.004
Tenure + Work + Age + Type	29.67	29	0.00	1	1.00

In fitting these models, work is entered as the two dummy variables work1 and work2, and similarly age is entered as age1 and age2.

---

and Figure 10.2(b) shows a normal probability plot of the Pearson residuals. These plots give no obvious cause for concern, so we can now try to interpret our fitted model. It appears that conditional on work and age, the probability of a positive response is far greater for respondents who own their home as opposed to those who rent. And conditional on age and tenure, unskilled and office workers tend to have a lower probability of responding yes than skilled workers. Finally, it appears that the two younger age groups do not differ in their probability of giving a positive response, and that this probability is greater than that in the oldest age group.

10.5. THE GENERALIZED LINEAR MODEL

In Chapter 6 we showed that the models used in the analysis of variance and those used in multiple linear regression are equivalent versions of a linear model in which an observed response is expressed as a linear function of explanatory variables plus some random disturbance term, often referred to as the "error" even though in many cases it may not have anything to do with measurement error. Therefore, the general form of such models, as outlined in several previous chapters, is

$$\text{observed response} = \text{expected response} + \text{error,} \qquad (10.11)$$

$$\text{expected response} = \text{linear function of explanatory variables.}$$

$$(10.12)$$

TABLE 10.19

Parameter Estimates and Standard Errors for the Final Model
Selected for Danish Do-It-Yourself Data

Parameter	Estimate	SE	t
Intercept	0.304	0.135	2.262
Tenure[a]	1.014	0.114	8.868
Work1	−0.763	0.152	−5.019
Work2	−0.305	0.141	−2.169
Age1	−0.113	0.137	−0.826
Age2	−0.436	0.140	−3.116

Observed and Predicted Probabilities of a Yes Response

Cell	Observed	Predicted	n
1	0.54	0.58	33
2	0.54	0.55	28
3	0.40	0.47	15
4	0.55	0.58	62
5	0.71	0.55	14
6	0.25	0.47	8
7	0.83	0.79	6
8	0.75	0.77	4
9	0.50	0.71	2
10	0.82	0.79	68
11	0.73	0.77	77
12	0.81	0.71	43
13	0.33	0.39	51
14	0.37	0.36	27
15	0.44	0.29	34
16	0.40	0.39	73
17	0.19	0.36	16
18	0.30	0.29	23
19	0.40	0.64	5
20	0.00	0.61	2
21	1.00	0.53	3
22	0.72	0.64	32
23	0.63	0.61	83
24	0.49	0.53	100
25	0.55	0.50	55
26	0.55	0.47	42
27	0.34	0.39	61
28	0.47	0.50	47
29	0.45	0.47	29
30	0.48	0.39	23

(Continued)

TABLE 10.19
(Continued)

Cell	Observed	Predicted	n
31	0.67	0.73	12
32	0.71	0.71	7
33	0.33	0.64	3
34	0.74	0.73	73
35	0.72	0.71	267
36	0.63	0.71	163

[a] Tenure is coded 0 for rent and 1 for own.

Display 10.6
Diagnostics for Logistic Regression

- The first diagnostics for a logistic regression are the *Pearson residuals* defined as

$$X_i = \frac{y_i - n_i \hat{p}_i}{\sqrt{n_i \hat{p}_i (1 - \hat{p}_i)}}.$$

- For $n_i \geq 5$ the distribution of the Pearson residuals can be reasonably approximated by a standard normal distribution, so a normal probability plot of the X_i should be linear.
- Pearson residuals with absolute values greater than two or three might be regarded with suspicion.
- The second useful diagnostic for checking a logistic regression model is the *deviance residual*, defined as

$$d_i = \text{sgn}(y_i - \hat{y}_i) \left[2y_i \ln \left(\frac{y_i}{\hat{y}_i} \right) + 2(n_1 - y_i) \ln \left(\frac{n_i - y_i}{n_i - \hat{y}_i} \right) \right] 1/2,$$

where $\text{sgn}(y_i - \hat{y}_i)$ is the function that makes d_i positive when $y_i \geq \hat{y}_i$ and negative when $y_i < \hat{y}_i$. Plotting the deviance residuals against the fitted values can be useful, indicating problems with models. Like the Pearson residuals, the deviance residual is approximately normally distributed.

Specification of the model is completed by assuming some specific distribution for the error terms. In the case of ANOVA and multiple regression models, for example, the assumed distribution is normal with mean zero and a constant variance σ^2.

Now consider the log-linear and the logistic regression models introduced in this chapter. How might these models be put into a similar form to the models used in the analysis of variance and multiple regression? The answer is by a relatively simple

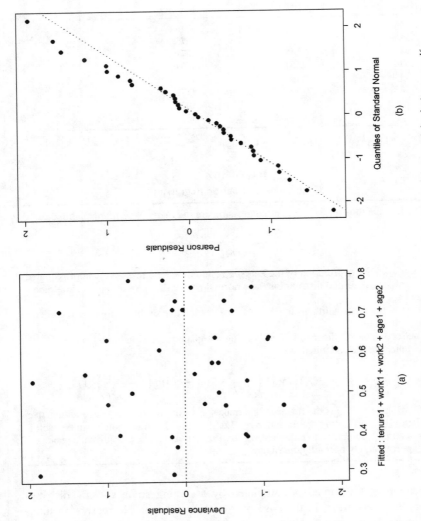

FIG. 10.2. Residual plots for the final model selected for Danish do-it-yourself data: (a) deviance residuals plotted against fitted values; (b) normal probability plot of Pearson residuals.

adjustment of the equations given above, namely allowing some transformation of the expected response to be modeled as a linear function of explanatory variables; that is, by introducing a model of the form

$$\text{observed response} = \text{expected response} + \text{error}, \qquad (10.13)$$

$$f(\text{expected response}) = \text{linear function of explanatory variables}, \qquad (10.14)$$

where f represents some suitable transformation. In the context of this *generalized linear model* (GLM), f is known as a *link function*. By also allowing the error terms to have distributions other than the normal, both log-linear models and logistic regression models can be included in the same framewok as ANOVA and multiple regression models. For example, for logistic regression, the link function would be the logistic and the error term binomial.

The GLM allowing a variety of link functions and numerous error distributions was first introduced into statistics by Nelder and Wedderburn (1972). Such models are fitted to data by using a general maximum likelihood approach, details of which are well outside the technical requirements of this text (see McCullagh and Nelder, 1989, for details). Apart from the unifying perspective of the GLM, its main advantage is that it provides the opportunity to carry out analyses that make more realistic assumptions about data than the normality assumption made explicitly and, more worryingly, often implicitly in the past. Nowadays, all statistical software packages can fit GLMs routinely, and researchers in general, and those working in psychology in particular, need to be aware of the possibilities such models often offer for a richer and more satisfactory analysis of their data.

10.6. SUMMARY

1. The analysis of cross-classifications of three or more categorical variables can now be undertaken routinely by using log-linear models.
2. The log-linear models fitted to multidimensional tables correspond to particular hypotheses about the variables forming the tables.
3. Expected values and parameters in log-linear models are estimated by maximum likelihood methods. In some cases the former consist of simple functions of particular marginal totals of observed frequencies. However, in many examples, the estimated frequencies have to be obtained by an iterative process.
4. The fit of a log-linear model is assessed by comparing the observed and estimated expected values under the model by means of the chi-squared statistic or, more commonly, the likelihood ratio statistic.
5. In a data set in which one of the categorical variables can be considered to be the response variable, logistic regression can be applied to investigate the

effects of explanatory variables. The regression parameters in such models can be interpreted in terms of odds ratios.

6. Categorical response variables with more than two categories and ordinal response variables can also be handled by logistic regression types of models; see Agresti (1996) for details.

COMPUTER HINTS

SPSS

The type of analyes described in this chapter can be accessed from the Statistics menu; for example, to undertake a logistic regression by using forward selection to choose a subset of explanatory variables, we can use the following.

1. Click on **Statistics**, click on **Regression**, and then click on **Logistic**.
2. Move the relevant binary dependent variable into the **Dependent** variable box.
3. Move the relevant explanatory variables into the **Covariates** box.
4. Choose **Forward** as the selected **Methods**.
5. Click on **OK**.

S-PLUS

In S-PLUS, log-linear analysis and logistic regression can be accessed by means of the Statistics menu, as follows.

1. Click on **Statistics**, click on **Regression**, and then click on **Log-linear (Poisson)** for the log-linear models dialog box, or **Logistic** for the logistic regression dialog box.
2. Click on the **Plot** tag to specify residual plots, and so on.

When the S-PLUS command line language is used, log-linear models and logistic regression models can be applied by using the *glm* function, specifying *family=poisson* for the former, and *family=binomial* for the later.

When the data for logistic regression are grouped as in the Danish do-it-yourself example in the text, then the total number of observations in each group has to be passed to the *glm* function by using the *weights* argument. So if, for example, the GHQ data were stored in a data frame *GHQ* with variables *sex* and *score*, the logistic regression command would be

$$glm(p \sim sex + score, family = binomial, weights = n, data = GHQ),$$

where p is the vector of observed proportions, and n is the vector of number of observations.

EXERCISES

10.1. The data in Table 10.20 were obtained from a study of the relationship between car size and car accident injuries. Accidents were classified according to their type, severity, and whether or not the driver was ejected. Using severity as the response variable, derive and interpret a suitable logistic model for these accounts.

10.2. The data shown in Table 10.21 arise from a study in which a sample of 1008 people were asked to compare two detergents, brand M and brand X. In addition to stating their preference, the sample members provided information on previous use of brand M, the degree of softness of the water that they used, and the temperature of the water. Use log-linear models to explore the associations between the four variables.

10.3. Show that the marginal totals of estimated expected values $E_{ij.}$, $E_{i.k}$, and $E_{.jk}$, corresponding to the no second-order relationship hypothesis for the suicide data, are equal to the corresponding marginal totals of the observed values.

10.4. The data in Table 10.22 (taken from Johnson and Albert, 1999) are for 30 students in a statistics class. The response variable y indicates whether

TABLE 10.20
Car Accident Data

| Car Weight | Driver Ejected | Accident Type | Number Hurt | |
			Severely	Not Severely
Small	No	Collision	150	350
Small	No	Rollover	112	60
Small	Yes	Collision	23	26
Small	Yes	Rollover	80	19
Standard	No	Collision	1022	1878
Standard	No	Rollover	404	148
Standard	Yes	Collision	161	111
Standard	Yes	Rollover	265	22

TABLE 10.21
Comparisons of Detergents Data

Water Softness	Brand Preferred	Previous User of M		Not Previous User of M	
		High Temp.	Low Temp.	High Temp.	Low Temp.
Soft	X	19	57	29	63
	M	29	49	27	53
Medium	X	23	47	33	66
	M	47	55	23	50
Hard	X	24	37	42	68
	M	43	52	30	42

TABLE 10.22
Data for Class of Statistics Students

Student	y_i	Test Score	Grade in Course
1	0	525	B
2	0	533	C
3	1	545	B
4	0	582	A
5	1	581	C
6	1	576	D
7	1	572	B
8	1	609	A
9	1	559	C
10	1	543	D
11	1	576	B
12	1	525	A
13	1	574	D
14	1	582	C
15	1	574	B
16	0	471	B
17	1	595	C
18	0	557	A
19	0	557	A
20	1	584	A

TABLE 10.23
Menstruation of Girls in Warsaw

x	y	n
11.08	2	120
11.33	5	88
11.58	10	105
11.83	17	111
12.08	16	100
12.33	29	93
12.58	39	100
12.83	51	108
13.08	47	99
13.33	67	106
13.58	81	117
14.08	79	98
14.33	90	97
14.58	93	100
15.08	117	122
15.33	107	111
15.58	92	94
17.58	1049	1049

Note. Numbers y who have reached menarche from n in age group with center x.

or not the student passed ($y = 1$) or failed ($y = 0$) the statistics examination at the end of the course. Also given are the students' scores on a previous math test and their grades for a prerequisite probability course.

1. Group the students into those with maths test scores <500, 501–550, 551–600, and >600, and then fit a linear model to the probability of passing by using the midpoint of the grouping interval as the explanatory variable.
2. Use your fitted model to predict the probability of passing for students with maths scores of 350 and 800.
3. Now fit a linear logistic model to the same data and again use the model to predict the probability of passing for students with math scores of 350 and 800.
4. Finally fit a logistic regression model to the ungrouped data by using both explanatory variables.

10.5. Show that the interaction parameter in the saturated log-linear model for a 2×2 contingency table is related to the odds ratio of the table.

10.6. The data in Table 10.23 relate to a sample of girls in Warsaw, the response variable indicating whether or not the girl has begun menstruation and the exploratory variable age in years (measured to the month). Plot the estimated probability of menstruation as a function of age and show the linear and logistic regression fits to the data on the plot.

10.7. Fit a logistic regression model to the GHQ data that includes main effects for both gender and GHQ and an interaction between the two variables.

10.8. Examine both the death penalty data (Table 10.3) and the infant's survival data (Table 10.4) by fitting suitable log-linear models and suggest what it is that leads to the spurious results when the data are aggregated over a particular variable.

Appendix A

Statistical Glossary

This glossary includes terms encountered in introductory statistics courses and terms mentioned in this text with little or no explanation. In addition, some terms of general statistical interest that are not specific to either this text or psychology are defined. Terms that are explained in detail in the text are *not* included in this glossary. Terms in italics in a definition are themselves defined in the appropriate place in the glossary. Terms are listed alphabetically, using the letter-by-letter convention. Four dictionaries of statistics that readers may also find useful are as follows.

1. Everitt, B. S. (1995). *The cambridge dictionary of statistics in the medical sciences*. Cambridge: Cambridge University Press.
2. Freund, J. E., and Williams, F. J. (1966). *Dictionary/outline of basic statistics*. New York: Dover.
3. Everitt, B. S. (1998). *The cambridge dictionary of statistics*. Cambridge: Cambridge University Press.
4. Everitt, B. S., and Wykes, T. (1999). *A dictionary of statistics for psychologists*. London: Arnold.

327

A

Acceptance region: The set of values of a *test statistic* for which the *null hypothesis* is accepted. Suppose, for example, a *z test* is being used to test that the mean of a population is 10 against the alternative that it is not 10. If the *significance level* chosen is .05, then the acceptance region consists of values of z between -1.96 and 1.96.

Additive effect: A term used when the effect of administering two treatments together is the sum of their separate effects. See also *additive model*.

Additive model: A model in which the explanatory variables have an *additive effect* on the response variable. So, for example, if variable A has an effect of size a on some response measure and variable B one of size b on the same response, then in an assumed additive model for A and B, their combined effect would be $a + b$.

Alpha(α):   The probability of a *Type I error*. See also *significance level*.

Alternative hypothesis: The hypothesis against which the *null hypothesis* is tested.

Analysis of variance: The separation of variance attributable to one cause from the variance attributable to others. Provides a way of testing for differences between a set of more than two population means.

A posteriori comparisons: Synonym for *post hoc comparisons*.

A priori comparisons: Synonym for *planned comparisons*.

Asymmetrical distribution: A *probability distribution* or *frequency distribution* that is not symmetrical about some central value. An example would be a distribution with positive *skewness* as shown in Figure A.1, giving the histogram of 200 reaction times (seconds) to a particular task.

Attenuation: A term applied to the correlation between two variables when both are subject to measurement error, to indicate that the value of the correlation between the true values is likely to be underestimated.

B

Balanced design: A term applied to any experimental design in which the same number of observations is taken for each combination of the experimental factors.

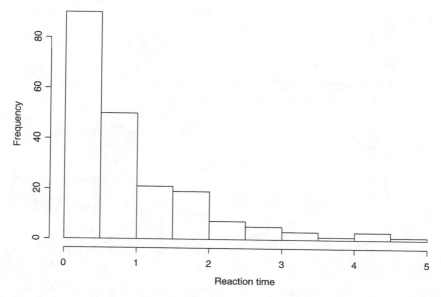

FIG. A.1.   Example of an asymmetrical distribution.

Bartlett's test: A test for the equality of the variances of a number of populations, sometimes used prior to applying *analysis of variance* techniques to assess the assumption of homogeneity of variance. It is of limited practical value because of its known sensitivity to nonnormality, so that a significant result might be caused by departures from normality rather than by different variances. See also *Box's test* and *Hartley's test*.

Bell-shaped distribution: A *probability distribution* having the overall shape of a vertical cross section of a bell. The *normal distribution* is the most well-known example, but a *Student's t distribution* is also this shape.

Beta coefficient: A regression coefficient that is standardized so as to allow for a direct comparison between explanatory variables as to their relative explanatory power for the response variable. It is calculated from the raw regression coefficients by multiplying them by the standard deviation of the corresponding explanatory variable and then dividing by the standard deviation of the response variable.

Bias: Deviation of results or inferences from the truth, or processes leading to such deviation. More specifically, this is the extent to which the statistical method used in a study does not estimate the quantity thought to be estimated.

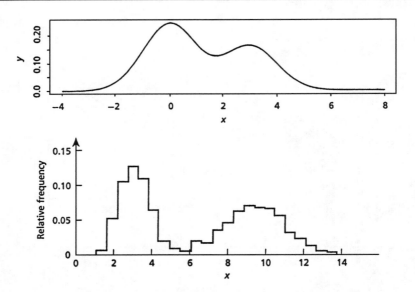

FIG. A.2.   Bimodal probability and frequency distributions.

Bimodal distribution: A *probability distribution*, or a *frequency distribution*, with two modes. Figure A.2 shows examples.

Binary variable: Observations that occur in one of two possible states, which are often labeled 0 and 1. Such data are frequently encountered in psychological investigations; commonly occurring examples include "improved/not improved," and "depressed/not depressed."

Binomial distribution: The *probability distribution* of the number of successes, x, in a series of n independent trials, each of which can result in either a success or failure. The probability of a success, p, remains constant from trial to trial. Specifically, the distribution of x is given by

$$P(x) = \frac{n!}{x!(n-x)!} p^x (1-p)^{n-x}, \quad x = 0, 1, 2, \ldots, n.$$

The mean of the distribution is *np* and its variance is *np*(1 − *p*).

Biserial correlation: A measure of the strength of the relationship between two variables, one continuous (y) and the other recorded as a *binary variable* (x), but having underlying continuity and normality. It is estimated from the sample values as

$$r_b = [(\bar{y}_1 - \bar{y}_0)/s_y]\,(pq/u),$$

where $\bar{y}_1$ is the sample mean of the y variable for those individuals for whom $x = 1$, $\bar{y}_0$ is the sample mean of the y variable for those individuals having $x = 0$, s_y is the standard deviation of the y values, p is the proportion of individuals with $x = 1$, and $q = 1 - p$ is the proportion of individuals with $x = 0$. Finally, u is the ordinate (height) of a *normal distribution* with mean zero and standard deviation one, at the point of division between the p and q proportions of the curve. See also *point-biserial correlation*.

Bivariate data: Data in which the subjects each have measurements on two variables.

Box's test: A test for assessing the equality of the variances in a number of populations that is less sensitive to departures from normality than *Bartlett's test*. See also *Hartley's test*.

C

Ceiling effect: A term used to describe what happens when many subjects in a study have scores on a variable that are at or near the possible upper limit (ceiling). Such an effect may cause problems for some types of analysis because it reduces the possible amount of variation in the variable. The converse, or *floor effect*, causes similar problems.

Central tendency: A property of the distribution of a variable usually measured by statistics such as the mean, median, and mode.

Change scores: Scores obtained by subtracting a posttreatment score on some variable from the corresponding pretreatment, baseline value.

Chi-squared distribution: The *probability distribution* of the sum of squares of a number of independent normal variables with means zero and standard deviations one. This distribution arises in many area of statistics, for example, assessing the goodness-of-fit of models, particularly those fitted to contingency tables.

Coefficient of determination: The square of the *correlation coefficient* between two variables x and y. Gives the proportion of the variation in one variable that is accounted for by the other. For example, a correlation of 0.8 implies that 64% of the variance of y is accounted for by x.

Coefficient of variation: A measure of spread for a set of data defined as

$$100 \times \text{standard deviation/mean.}$$

This was originally proposed as a way of comparing the variability in different distributions, but it was found to be sensitive to errors in the mean.

Commensurate variables: Variables that are on the same scale or expressed in the same units, for example, systolic and diastolic blood pressure.

Composite hypothesis: A hypothesis that specifies more than a single value for a parameter, for example, the hypothesis that the mean of a population is greater than some value.

Compound symmetry: The property possessed by a *covariance matrix* of a set of multivariate data when its main diagonal elements are equal to one another, and additionally its off-diagonal elements are also equal. Consequently, the matrix has the general form

$$
\Sigma = \begin{pmatrix}
\sigma^2 & \rho\sigma^2 & \cdots & \rho\sigma^2 \\
\rho\sigma^2 & \sigma^2 & \cdots & \rho\sigma^2 \\
\vdots & \vdots & & \vdots \\
\rho\sigma^2 & \rho\sigma^2 & \cdots & \rho\sigma^2
\end{pmatrix},
$$

where ρ is the assumed common *correlation coefficient* of the measures.

Confidence interval: A range of values, calculated from the sample observations, that are believed, with a particular probability, to contain the true parameter value. A 95% confidence interval, for example, implies that if the estimation process were repeated again and again, then 95% of the calculated intervals would be expected to contain the true parameter value. Note that the stated probability level refers to properties of the interval and not to the parameter itself, which is not considered a *random variable*.

Conservative and nonconservative tests: Terms usually encountered in discussions of *multiple comparison tests*. Nonconservative tests provide poor control over the *per-experiment error rate*. Conservative tests, in contrast, may limit the *per-comparison error rate* to unecessarily low values, and tend to have low *power* unless the sample size is large.

Contrast: A linear function of parameters or statistics in which the coefficients sum to zero. It is most often encountered in the context of analysis of variance. For example, in an application involving, say, three treatment groups (with means x_{T_1}, x_{T_2}, and x_{T_3}) and a control group (with mean x_C), the following is the contrast for comparing the mean of the control group to the average of the treatment groups:

$$
x_C - 1/3x_{T_1} - 1/3x_{T_2} - 1/3x_{T_3}.
$$

See also *orthogonal contrast*.

Correlation coefficient: An index that quantifies the linear relationship between a pair of variables. For sample observations a variety of such coefficients have been suggested, of which the most commonly used is *Pearson's product moment correlation coefficient*, defined as

$$r = \frac{\sum_{i=1}^{n}(x_i - \bar{x})(y_i - \bar{y})}{\sqrt{\sum_{i=1}^{n}(x_i - \bar{x})^2(y_i - \bar{y})^2}},$$

where $(x_1, y_1), (x_2, y_2), \ldots, (x_n, y_n)$ are the n sample values of the two variables of interest. The coefficient takes values between -1 and 1, with the sign indicating the direction of the relationship and the numerical magnitude its strength. Values of -1 or 1 indicate that the sample values fall on a straight line. A value of zero indicates the lack of any linear relationship between the two variables.

Correlation matrix: A square, *symmetric matrix* with rows and columns corresponding to variables, in which the off-diagonal elements are the *correlation coefficients* between pairs of variables, and elements on the main diagonal are unity.

Covariance: For a sample of n pairs of observation, $(x_1, y_1), (x_2, y_2), \ldots, (x_n, y_n)$, the statistic given by

$$c_{xy} = \frac{1}{n}\sum_{i=1}^{n}(x_i - \bar{x})(y_i - \bar{y}),$$

where $\bar{x}$ and $\bar{y}$ are the sample means of the x_i and y_i respectively.

Covariance matrix: A *symmetric matrix* in which the off-diagonal elements are the *covariances* of pairs of variables, and the elements on the main diagonal are variances.

Critical region: The values of a *test statistic* that lead to rejection of a *null hypothesis*. The size of the critical region is the probability of obtaining an outcome belonging to this region when the null hypothesis is true, that is, the probability of a *Type I error*. See also *acceptance region*.

Critical value: The value with which a statistic calculated form the sample data is compared in order to decide whether a *null hypothesis* should be rejected. The value is related to the particular significance level chosen.

Cronbach's alpha: An index of the internal consistency of a psychological test. If the test consists of n items and an individual's score is the total answered correctly, then the coefficient is given specifically by

$$\alpha = \frac{n}{n-1}\left(1 - \frac{1}{\sigma^2}\sum_{i=1}^{n}\sigma_i^2\right),$$

where σ^2 is the variance of the total scores and σ_i^2 is the variance of the set of 0,1 scores representing correct and incorrect answers on item i.

Cross-validation: The division of data into two approximately equal sized subsets, one of which is used to estimate the parameters in some model of interest, and the other to assess whether the model with these parameter values fits adequately.

Cumulative frequency distribution: A listing of the sample values of a variable, together with the proportion of the observations less than or equal to each value.

D

Data dredging: A term used to describe comparisons made within a data set not specifically prescribed prior to the start of the study.

Data reduction: The process of summarizing large amounts of data by forming *frequency distributions, histograms, scatter diagrams*, and so on, and calculating statistics such as means, variances, and correlation coefficients. The term is also used when a low-dimensional representation of multivariate data is sought by use of procedures such as principal components analysis and factor analysis.

Data set: A general term for observations and measurements collected during any type of scientific investigation.

Degrees of freedom: An elusive concept that occurs throughout statistics. Essentially, the term means the number of independent units of information in a sample relevant to the estimation of a parameter or calculation of a statistic. For example, in a 2×2 contingency table with a given set of marginal totals, only one of the four cell frequencies is free and the table has therefore a single degree of freedom. In many cases the term corresponds to the number of parameters in a model.

Dependent variable: See *response variable*.

Descriptive statistics: A general term for methods of summarizing and tabulating data that make their main features more transparent, for example, calculating means and variances and plotting histograms. See also *exploratory data analysis* and *initial data analysis*.

DF(df):   Abbreviations for *degrees of freedom*.

Diagonal matrix: A *square matrix* whose off-diagonal elements are all zero. For example,

$$\mathbf{D} = \begin{pmatrix} 10 & 0 & 0 \\ 0 & 5 & 0 \\ 0 & 0 & 3 \end{pmatrix}.$$

Dichotomous variable: Synonym for *binary variable*.

Digit preference: The personal and often subconscious *bias* that frequently occurs in the recording of observations. It is usually most obvious in the final recorded digit of a measurement.

Discrete variables: Variables having only integer values, for example, number of trials to learn a particular task.

Doubly multivariate data: A term used for the data collected in those longitudinal studies in which more than a single response variable is recorded for each subject on each occasion.

Dummy variables: The variables resulting from recoding *categorical variables* with more than two categories into a series of *binary variables*. Marital status, for example, if originally labeled 1 for married; 2 for single; and 3 for divorced, widowed, or separated could be redefined in terms of two variables as follows.

Variable 1:   1 if single, and 0 otherwise;
Variable 2:   1 if divorced, widowed, or separated, and 0 otherwise.

For a married person, both new variables would be zero. In general a categorical variable with k categories would be recoded in terms of $k-1$ dummy variables. Such recoding is used before polychotomous variables are used as explanatory variables in a regression analysis to avoid the unreasonable assumption that the original numerical codes for the categories, that is, the values $1, 2, \ldots, k$, correspond to an *interval scale*.

E

EDA: Abbreviation for *exploratory data analysis*.

Effect: Generally used for the change in a response variable produced by a change in one or more explanatory or factor variables.

Empirical: Based on observation or experiment rather than deduction from basic laws or theory.

Error rate: The proportion of subjects misclassified by an allocation rule derived from a discriminant analysis.

Estimation: The process of providing a numerical value for a population parameter on the basis of information collected from a sample. If a single figure is calculated for the unknown parameter, the process is called *point estimation*. If an interval is calculated within which the parameter is likely to fall, then the procedure is called *interval estimation*. See also *least-squares estimation* and *confidence interval*.

Estimator: A statistic used to provide an estimate for a parameter. The sample mean, for example, is an *unbiased* estimator of the population mean.

Experimental design: The arrangement and procedures used in an *experimental study*. Some general principles of good design are simplicity, avoidance of *bias*, the use of *random allocation* for forming treatment groups, replication, and adequate sample size.

Experimental study: A general term for investigations in which the researcher can deliberately influence events and investigate the effects of the intervention.

Experimentwise error rate: Synonym for *per-experiment error rate*.

Explanatory variables: The variables appearing on the right-hand side of the equations defining, for example, multiple regression or logistic regression, and that seek to predict or explain the response variable. Also commonly known as the independent variables, although this is not to be recommended because they are rarely independent of one another.

Exploratory data analysis: An approach to data analysis that emphasizes the use of informal graphical procedures not based on prior assumptions about the structure of the data or on formal models for the data. The essence of this approach is that, broadly speaking, data are assumed to possess the following structure:

$$\text{Data} = \text{Smooth} + \text{Rough},$$

where Smooth is the underlying regularity or pattern in the data. The objective of the exploratory approach is to separate the Smooth from the Rough with minimal use of formal mathematics or statistical methods. See also *initial data analysis*.

Eyeball test: Informal assessment of data simply by inspection and mental calculation allied with experience of the particular area from which the data arise.

F

Factor: A term used in a variety of ways in statistics, but most commonly to refer to a categorical variable, with a small number of levels, under investigation in an experiment as a possible source of variation. This is essentially simply a categorical explanatory variable.

Familywise error rate: The probability of making any error in a given family of inferences. See also per-comparison error rate and *per-experiment error rate*.

F distribution: The *probability distribution* of the ratio of two independent *random variables*, each having a *chi-squared distribution*. Divided by their respective degrees of freedom.

Fisher's exact test: An alternative procedure to the use of the chi-squared statistic for assessing the independence of two variables forming a 2×2 contingency table, particularly when the expected frequencies are small.

Fisher's z transformation: A transformation of Pearson's product moment correlation coefficient, r, given by

$$z = \frac{1}{2} \ln \frac{1+r}{1-r}.$$

The statistic z has mean $\frac{1}{2} \ln(1+\rho)/(1-\rho)$, where ρ is the population correlation value and variance $1/(n-3)$ where n is the sample size. The transformation may be used to test hypotheses and to construct *confidence intervals* for ρ.

Fishing expedition: Synonym for *data dredging*.

Fitted value: Usually used to refer to the value of the response variable as predicted by some estimated model.

Follow-up: The process of locating research subjects or patients to determine whether or not some outcome of interest has occurred.

Floor effect: See *ceiling effect*.

Frequency distribution: The division of a sample of observations into a number of classes, together with the number of observations in each class. Acts as a useful summary of the main features of the data such as location, shape, and spread. An example of such a table is given below.

	IQ Scores
Class Limits	*Observed Frequency*
75–79	1
80–84	2
85–89	5
90–94	9
95–99	10
100–104	7
105–109	4
110–114	2
≥115	1

Frequency polygon: A diagram used to display graphically the values in a *frequency distribution*. The frequencies are graphed as ordinate against the class midpoints as abcissae. The points are then joined by a series of straight lines. Particularly useful in displaying a number of frequency distributions on the same diagram.

***F* test:**   A test for the equality of the variances of two populations having *normal distributions*, based on the ratio of the variances of a sample of observations taken from each. It is most often encountered in the analysis of variance, in which testing whether particular variances are the same also tests for the equality of a set of means.

G

Gambler's fallacy: The belief that if an event has not happened for a long time, it is bound to occur soon.

Goodness-of-fit statistics: Measures of agreement between a set of sample values and the corresponding values predicted from some model of interest.

Grand mean: Mean of all the values in a grouped data set irrespective of groups.

Graphical methods: A generic term for those techniques in which the results are given in the form of a graph, diagram, or some other form of visual display.

H

H_0:   Symbol for *null hypothesis*.

H_1:   Symbol for *alternative hypothesis*.

Halo effect: The tendency of a subject's performance on some task to be over-rated because of the observer's perception of the subject "doing well" gained in an earlier exercise or when assessed in a different area.

Harmonic mean: The reciprocal of the arithmetic mean of the reciprocals of a set of observations, $x_1, x_2, \ldots, x_n$. Specifically obtained from

$$\frac{1}{H} = \frac{1}{n} \sum_{i=1}^{n} \frac{1}{x_i}.$$

Hartley's test: A simple test of the equality of variances of a number of populations. The *test statistic* is the ratio of the largest to the smallest sample variances.

Hawthorne effect: A term used for the effect that might be produced in an experiment simply from the awareness by the subjects that they are participating in some form of scientific investigation. The name comes from a study of industrial efficiency at the Hawthorne Plant in Chicago in the 1920s.

Hello–goodbye effect: A phenomenon originally described in psychotherapy research but one that may arise whenever a subject is assessed on two occasions with some intervention between the visits. Before an intervention, a person may present himself or herself in as bad as light as possible, thereby hoping to qualify for treatment, and impressing staff with the seriousness of his or her problems. At the end of the study the person may want to please the staff with his or her improvement, and so may minimize any problems. The result is to make it appear that there has been some impovement when none has occurred, or to magnify the effects that did occur.

Histogram: A graphical representation of a set of observations in which class frequencies are represented by the areas of rectangles centered on the class interval. If the latter are all equal, the heights of the rectangles are also proportional to the observed frequencies.

Homogeneous: A term that is used in statistics to indicate the equality of some quantity of interest (most often a variance), in a number of different groups, populations, and so on.

Hypothesis testing: A general term for the procedure of assessing whether sample data are consistent or otherwise with statements made about the population. See also *null hypothesis, alternative hypothesis, composite hypothesis, significance test, significance level, Type I error*, and *Type II error*.

I

IDA: Abbreviation for *initial data analysis*.

Identification: The degree to which there is sufficient information in the sample observations to estimate the parameters in a proposed model. An unidentified model is one in which there are too many parameters in relation to the number of observations to make estimation possible. A just identified model corresponds to a saturated model. Finally, an overidentified model is one in which parameters can be estimated, and there remain degrees of freedom to allow the fit of the model to be assessed.

Identity matrix: A *diagonal matrix* in which all the elements on the leading diagonal are unity and all the other elements are zero.

Independence: Essentially, two events are said to be independent if knowing the outcome of one tells us nothing about the other. More formally the concept is defined in terms of the probabilities of the two events. In particular, two events A and B are said to be independent if

$$P(A \text{ and } B) = P(A) \times P(B),$$

where $P(A)$ and $P(B)$ represent the probabilities of A and B.

Independent samples t test: See *Student's t test*.

Inference: The process of drawing conclusions about a population on the basis of measurements or observations made on a sample of individuals from the population.

Initial data analysis: The first phase in the examination of a data set, which consists of a number of informal steps, including

• checking the quality of the data,

- calculating simple summary statistics, and
- constructing appropriate graphs.

The general aim is to clarify the structure of the data, obtain a simple descriptive summary, and perhaps get ideas for a more sophisticated analysis.

Interaction: A term applied when two (or more) explanatory variables do not act independently on a response variable. See also *additive effect*.

Interval variable: Synonym for continuous variable.

Interval estimation: See *estimation*.

Interval variable: Synonym for *continuous variable*.

Interviewer bias: The *bias* that may occur in surveys of human populations because of the direct result of the action of the interviewer. The bias can arise for a variety of reasons, including failure to contact the right persons and systematic errors in recording the answers received from the respondent.

J

J-shaped distribution: An extremely assymetrical distribution with its maximum frequency in the initial class and a declining frequency elsewhere. An example is shown in Figure A.3.

K

Kurtosis: The extent to which the peak of a unimodal *frequency distribution* departs from the shape of a *normal distribution*, by either being more pointed (leptokurtic) or flatter (platykurtic). It is usually measured for a probability distribution as

$$\mu_4/\mu_2^2 - 3,$$

where μ_4 is the fourth central moment of the distribution, and μ_2 is its variance. (Corresponding functions of the sample moments are used for frequency distributions.) For a normal distribution, this index takes the value zero (other distributions with zero kurtosis are called mesokurtic); for one that is leptokurtic it is positive, and for a platykurtic curve it is negative.

L

Large sample method: Any statistical method based on an approximation to a *normal distribution* or other *probability distribution* that becomes more accurate as sample size increases.

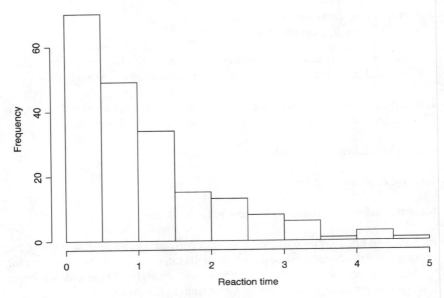

FIG. A.3.   Example of a J-shaped distribution.

Least-squares estimation: A method used for estimating parameters, particularly in regression analysis, by minimizing the difference between the observed response and the value predicted by the model. For example, if the expected value of a response variable y is of the form

$$E(y) = \alpha + \beta x,$$

where x is an explanatory variable, then least-squares estimators of the parameters α and β may be obtained from n pairs of sample values (x_1, y_1), (x_2, y_2), ..., (x_n, y_n) by minimizing S given by

$$S = \sum_{i=1}^{n}(y_i - \alpha - \beta x_i)^2,$$

to give

$$\hat{\alpha} = \bar{y} - \hat{\beta}\bar{x}$$

$$\hat{\beta} = \frac{\sum_{i=1}^{n}(x_i - \bar{x})(y_i - \bar{y})}{\sum_{i=1}^{n}(x_i - \bar{x})^2}.$$

Often referred to as ordinary least squares to differentiate this simple version of the technique from more involved versions, such as weighted least squares.

Leverage points: A term used in regression analysis for those observations that have an extreme value on one or more of the explanatory variables. The effect of such points is to force the fitted model close to the observed value of the response, leading to a small residual.

Likert scales: Scales often used in studies of attitudes, in which the raw scores are based on graded alternative responses to each of a series of questions. For example, the subject may be asked to indicate his or her degree of agreement with each of a series of statements relevant to the attitude. A number is attached to each possible response, for example, 1, strongly approve; 2, approve; 3, undecided; 4, dissaprove; 5, strongly disapprove. The sum of these is used as the composite score.

Logarithmic transformation: The transformation of a variable, x, obtained by taking $y = \ln(x)$. Often used when the *frequency distribution* of the variable, x, shows a moderate to large degree of *skewness* in order to achieve normality.

Lower triangular matrix: A matrix in which all the elements above the main diagonal are zero. An example is the following:

$$\mathbf{L} = \begin{pmatrix} 1 & 0 & 0 & 0 \\ 2 & 3 & 0 & 0 \\ 1 & 1 & 3 & 0 \\ 1 & 5 & 6 & 7 \end{pmatrix}.$$

M

Main effect: An estimate of the independent effect of usually a factor variable on a response variable in an ANOVA.

Manifest variable: A variable that can be measured directly, in contrast to a latent variable.

Marginal totals: A term often used for the total number of observations in each row and each column of a contingency table.

MANOVA: Acronym for *multivariate analysis of variance*.

Matched pairs: A term used for observations arising from either two individuals who are individually matched on a number of variables, for example, age and sex, or where two observations are taken on the same individual on two separate occasions. Essentially synonymous with *paired samples*.

Matched pairs t test: A *Student's t test* for the equality of the means of two populations, when the observations arise as *paired samples*. The test is based on the differences between the observations of the matched pairs. The *test statistic* is given by

$$t = \frac{\bar{d}}{s_d/\sqrt{n}},$$

where n is the sample size, $\bar{d}$ is the mean of the differences, and s_d is their standard deviation. If the null hypothesis of the equality of the population means is true, then t has a *Student's t distribution* with $n - 1$ degrees of freedom.

Matching: The process of making a study group and a comparison group comparable with respect to extraneous factors. It is often used in retrospective studies in the selection of cases and controls to control variation in a response variable that is due to sources other than those immediately under investigation. Several kinds of matching can be identified, the most common of which is when each case is individually matched with a control subject on the matching variables, such as age, sex, occupation, and so on. See also *paired samples*.

Matrix: A rectangular arrangement of numbers, algebraic functions, and so on. Two examples are

$$\mathbf{A} = \begin{pmatrix} 1 & 1 & 2 \\ 2 & 1 & 7 \end{pmatrix},$$

$$\mathbf{B} = \begin{pmatrix} b_{11} & b_{12} \\ b_{21} & b_{22} \\ b_{31} & b_{32} \\ b_{41} & b_{42} \end{pmatrix}.$$

Mean: A measure of location or central value for a continuous variable. For a sample of observations $x_1, x_2, \ldots, x_n$, the measure is calculated as

$$\bar{x} = \frac{\sum_{i=1}^{n} x_i}{n}$$

Mean vector: A vector containing the mean values of each variable in a set of multivariate data.

Measurement error: Errors in reading, calculating, or recording a numerical value. This is the difference between observed values of a variable recorded under similar conditions and some underlying true value.

Measures of association: Numerical indices quantifying the strength of the statistical dependence of two or more qualitative variables.

Median: The value in a set of ranked observations that divides the data into two parts of equal size. When there is an odd number of observations, the median is the middle value. When there is an even number of observations, the measure is calculated as the average of the two central values. It provides a measure of location of a sample that is suitable for *asymmetrical distributions* and is also relatively insensitive to the presence of *outliers*. See also *mean* and *mode*.

Misinterpretation of p values: A *p value* is commonly interpreted in a variety of ways that are incorrect. Most common is that it is the probability of the null hypothesis, and that it is the probability of the data having arisen by chance. For the correct interpretation, see the entry for *p value*.

Mixed data: Data containing a mixture of continuous variables, ordinal variables, and categorical variables.

Mode: The most frequently occurring value in a set of observations. Occasionally used as a measure of location. See also *mean* and *median*.

Model: A description of the assumed structure of a set of observations that can range from a fairly imprecise verbal account to, more usually, a formalized mathematical expression of the process assumed to have generated the observed data. The purpose of such a description is to aid in understanding the data.

Model building: A procedure that attempts to find the simplest model for a sample of observations that provides an adequate fit to the data.

Most powerful test: A test of a *null hypothesis* which has greater *power* than any other test for a given *alternative hypothesis*.

Multilevel models: Models for data that are organized hierachically, for example, children within families, that allow for the possibility that measurements made on children from the same family are likely to be correlated.

Multinomial distribution: A generalization of the *binomial distribution* to situations in which r outcomes can occur on each of n trials, where $r > 2$. Specifically the distribution is given by

$$P(n_1, n_2, \ldots, n_r) = \frac{n!}{n_1! n_2! \cdots n_r!} p_1^{n_1} p_2^{n_2} \cdots p_r^{n_r}$$

where n_i is the number of trials with outcome i, and p_i is the probability of outcome i occurring on a particular trial.

Multiple comparison tests: Procedures for detailed examination of the differences between a set of means, usually after a general hypothesis that they are all equal has been rejected. No single technique is best in all situations and a major distinction between techniques is how they control the possible inflation of the *type I error*.

Multivariate analysis: A generic term for the many methods of analysis important in investigating multivariate data.

Multivariate analysis of variance: A procedure for testing the equality of the *mean vectors* of more than two populations. The technique is directly analogous to the *analysis of variance* of *univariate data*, except that the groups are compared on q response variables simultaneously. In the univariate case, F tests are used to assess the hypotheses of interest. In the multivariate case, no single *test statistic* can be constructed that is optimal in all situations. The most widely used of the available test statistics is *Wilk's lambda*, which is based on three matrices $\mathbf{W}$ (the *within groups matrix of sums of squares and products*), $\mathbf{T}$ (the *total matrix of sums of squares and cross products*) and $\mathbf{B}$ (the *between groups matrix of sums of squares and cross products*), defined as follows:

$$\mathbf{T} = \sum_{i=1}^{g} \sum_{j=1}^{n_i} (\mathbf{x}_{ij} - \bar{\mathbf{x}})(\mathbf{x}_{ij} - \bar{\mathbf{x}})',$$

$$\mathbf{W} = \sum_{i=1}^{g} \sum_{j=1}^{n_i} (\mathbf{x}_{ij} - \bar{\mathbf{x}}_i)(\mathbf{x}_{ij} - \bar{\mathbf{x}}_i)',$$

$$\mathbf{B} = \sum_{i=1}^{g} n_i (\bar{\mathbf{x}}_i - \bar{\mathbf{x}})(\bar{\mathbf{x}}_i - \bar{\mathbf{x}})',$$

where $\mathbf{x}_{ij}, i = 1, \ldots, g, j = 1, \ldots, n_i$ represent the jth multivariate observation in the ith group, g is the number of groups, and n_i is the number of observations in the ith group. The mean vector of the ith group is represented by $\bar{\mathbf{x}}_i$ and the mean vector of all the observations by $\bar{\mathbf{x}}$. These matrices satisfy the equation

$$\mathbf{T} = \mathbf{W} + \mathbf{B}.$$

Wilk's lambda is given by the ratio of the determinants of $\mathbf{W}$ and $\mathbf{T}$, that is,

$$\Lambda = \frac{|\mathbf{W}|}{|\mathbf{T}|} = \frac{|\mathbf{W}|}{|\mathbf{W} + \mathbf{B}|}.$$

The statistic, Λ, can be transformed to give an F test to assess the null hypothesis of the equality of the population mean vectors. In addition to Λ, a number of other test statistics are available.

- Roy's largest root criterion: the largest eigenvalue of $\mathbf{BW}^{-1}$;
- The Hotelling–Lawley trace: the sum of the eigenvalues of $\mathbf{BW}^{-1}$;
- The Pillai–Bartlett trace: the sum of the eigenvalues of $\mathbf{BT}^{-1}$.

It has been found that the differences in *power* between the various test statistics are generally quite small, and so in most situations the choice will not greatly affect conclusions.

Multivariate normal distribution: The *probability distribution* of a set of variables $\mathbf{x}' = [x_1, x_2, \ldots, x_q]$ given by

$$f(x_1, x_2, \ldots, x_q) = (2\pi)^{-q/2} |\mathbf{\Sigma}|^{-1/2} \exp -\frac{1}{2}(\mathbf{x} - \boldsymbol{\mu})'\mathbf{\Sigma}^{-1}(\mathbf{x} - \boldsymbol{\mu}),$$

where $\boldsymbol{\mu}$ is the *mean vector* of the variables and $\mathbf{\Sigma}$ is their *variance–covariance matrix*. This distribution is assumed by multivariate analysis procedures such as *multivariate analysis of variance*.

N

Newman–Keuls test: A *multiple comparison test* used to investigate in more detail the differences between a set of means, as indicated by a significant F *test* in an analysis of variance.

Nominal significance level: The significance level of a test when its assumptions are valid.

Nonorthogonal designs: Analysis of variance designs with two or more factors in which the number of observations in each cell are not equal.

Normal distribution: A *probability distribution* of a *random variable*, x, that is assumed by many statistical methods. It is specifically given by

$$f(x) = \frac{1}{\sigma\sqrt{2\pi}} \exp\left[-\frac{1}{2}\frac{(x - \mu)^2}{\sigma^2}\right],$$

where μ and σ^2 are, respectively, the mean and variance of x. This distribution is bell shaped.

Null distribution: The *probability distribution* of a *test statistic* when the *null hypothesis* is true.

Null hypothesis: The "no difference" or "no association" hypothesis to be tested (usually by means of a significance test) against an *alternative hypothesis* that postulates a nonzero difference or association.

Null matrix: A matrix in which all elements are zero.

Null vector: A *vector*, the elements of which are all zero.

O

One-sided test: A significance test for which the *alternative hypothesis* is directional; for example, that one population mean is greater than another. The choice between a one-sided and *two-sided test* must be made before any *test statistic* is calculated.

Orthogonal: A term that occurs in several areas of statistics with different meanings in each case. It is most commonly encountered in relation to two variables or two linear functions of a set of variables to indicate statistical independence. It literally means at right angles.

Orthogonal contrasts: Sets of linear functions of either parameters or statistics in which the defining coefficients satisfy a particular relationship. Specifically, if c_1 and c_2 are two *contrasts* of a set of m parameters such that

$$c_1 = a_{11}\beta_1 + a_{12}\beta_2 + \cdots + a_{1m}\beta_m,$$
$$c_2 = a_{21}\beta_1 + a_{22}\beta_2 + \cdots + a_{2m}\beta_m,$$

they are orthogonal if $\sum_{i=1}^{m} a_{1i}a_{2i} = 0$. If, in addition, $\sum_{i=1}^{m} a_{1i}^2 = 1$ and $\sum_{i=1}^{m} a_{2i}^2 = 1$, then the contrasts are said to be orthonormal.

Orthogonal matrix: A *square matrix* that is such that multiplying the matrix by its transpose results in an *identity matrix*.

Outlier: An observation that appears to deviate markedly from the other members of the sample in which it occurs. In the set of systolic blood pressures, {125, 128, 130, 131, 198}, for example, 198 might be considered an outlier. Such extreme observations may be reflecting some abnormality in the measured characteristic of a patient, or they may result from an error in the measurement or recording.

P

Paired samples: Two samples of observations with the characteristic feature that each observation in one sample has one and only one matching observation in the other sample. There are several ways in which such samples can arise in psychological investigations. The first, self-pairing, occurs when each subject serves as his or her own control, as in, for example, therapeutic trials in which each subject receives both treatments, one on each of two separate occasions. Next, natural pairing can arise particularly, for example, in laboratory experiments involving litter-mate controls. Lastly, artificial pairing may be used by an investigator to match the two subjects in a pair on important characteristics likely to be related to the response variable.

Paired samples *t* test: Synonym for *matched pairs t test*.

Parameter: A numerical characteristic of a population or a model. The probability of a success in a *binomial distribution*, for example.

Partial correlation: The correlation between a pair of variables after adjusting for the effect of a third. It can be calculated from the sample correlation coefficients of each pair of variables involved as

$$r_{12|3} = \frac{r_{12} - r_{13}r_{23}}{\sqrt{\left(1 - r_{13}^2\right)\left(1 - r_{23}^2\right)}}.$$

Pearson's product moment correlation coefficient: An index that quantifies the linear relationship between a pair of variables. For a sample of n observations of two variables $(x_1, y_1), (x_2, y_2) \cdots (x_n, y_n)$ calculated as

$$r = \frac{\sum_{i=1}^{n}(x_i - \bar{x})(y_i - \bar{y})}{\sqrt{\sum_{i=1}^{n}(x_i - \bar{x})^2(y_i - \bar{y})^2}}.$$

The coefficient takes values between -1 and 1.

Per comparison error rate: The significance level at which each test or comparison is carried out in an experiment.

Per-experiment error rate: The probability of incorrectly rejecting at least one null hypothesis in an experiment involving one or more tests or comparisons, when the corresponding *null hypothesis* is true in each case. See also *per-comparison error rate*.

Placebo: A treatment designed to appear exactly like a comparison treatment, but which is devoid of the active component.

Planned comparisons: Comparisons between a set of means suggested before data are collected. Usually more powerful than a general test for mean differences.

Point-biserial correlation: A special case of *Pearson's product moment correlation coefficient* used when one variable is continuous (y) and the other is a *binary variable* (x) representing a natural dichotomy. Given by

$$r_{pb} = \frac{\bar{y}_1 - \bar{y}_0}{s_y}\sqrt{pq},$$

where $\bar{y}_1$ is the sample mean of the y variable for those individuals with $x = 1$, $\bar{y}_0$ is the sample mean of the y variable for those individuals with $x = 0$, s_y is the standard deviation of the y values, p is the proportion of individuals with $x = 1$, and $q = 1 - p$ is the proportion of individuals with $x = 0$. See also *biserial correlation*.

Poisson distribution: The *probability distribution* of the number of occurrences of some random event, x, in an interval of time or space, and given by

$$P(x) = \frac{e^{-\lambda}\lambda^x}{x!} \quad x = 0, 1, 2, \ldots.$$

The mean and variance of a variable with such a distribution are both equal to λ.

Population: In statistics this term is used for any finite or infinite collection of units, which are often people but may be, for example, institutions or events. See also *sample*.

Power: The probability of rejecting the *null hypothesis* when it is false. Power gives a method of discriminating between competing tests of the same hypothesis; the test with the higher power is preferred. It is also the basis of procedures for estimating the sample size needed to detect an effect of a particular magnitude.

Probability: The quantitative expression of the chance that an event will occur. This can be defined in a variety of ways, of which the most common is still that involving long-term relative frequency:

$$P(A) = \frac{\text{number of times } A \text{ occurs}}{\text{number of times } A \text{ could occur}}.$$

For example, if out of 100,000 children born in a region, 51,000 are boys, then the probability of a boy is 0.51.

Probability distribution: For a discrete *random variable*, this is a mathematical formula that gives the probability of each value of the variable. See, for example, *binomial distribution* and *Poisson distribution*. For a continous random variable, this is a curve described by a mathematical formula that specifies, by way of areas under the curve, the probability that the variable falls within a particular interval. An example is the *normal distribution*. In both cases the term probability density is also used. (A distinction is sometimes made between density and distribution, when the latter is reserved for the probability that the random variable will fall below some value.)

p **value:**   The probability of the observed data (or data showing a more extreme departure from the *null hypothesis*) when the null hypothesis is true. See also *misinterpretation of p values, significance test,* and *significance level.*

Q

Quasi-experiment: A term used for studies that resemble experiments but are weak on some of the characteristics, particularly that manipulation of subjects to groups is not under the investigator's control. For example, if interest centered on the health effects of a natural disaster, those who experience the disaster can be compared with those who do not, but subjects cannot be deliberately assigned (randomly or not) to the two groups. See also *experimental design.*

R

Randomization tests: Procedures for determining statistical significance directly from data without recourse to some particular *sampling distribution.* The data are divided (permuted) repeatedly between treatments, and for each division (permutation) the relevant *test statistic* (for example, a *t* or *F*) is calculated to determine the proportion of the data permutations that provide as large a test statistic as that associated with the observed data. If that proportion is smaller than some significance level α, the results are significant at the α level.

Random sample: Either a set of *n* independent and identically distributed *random variables*, or a sample of *n* individuals selected from a population in such a way that each sample of the same size is equally likely.

Random variable: A variable, the values of which occur according to some specified *probability distribution.*

Random variation: The variation in a data set unexplained by identifiable sources.

Range: The difference between the largest and smallest observations in a data set. This is often used as an easy-to-calculate measure of the dispersion in a set of observations, but it is not recommended for this task because of its sensitivity to *outliers.*

Ranking: The process of sorting a set of variable values into either ascending or descending order.

Rank of a matrix: The number of linearly independent rows or columns of a matrix of numbers.

Ranks: The relative positions of the members of a sample with respect to some characteristic.

Rank correlation coefficients: Correlation coefficients that depend on only the ranks of the variables, not on their observed values.

Reciprocal transformation: A transformation of the form $y = 1/x$, which is particularly useful for certain types of variables. Resistances, for example, become conductances, and times become speeds.

Regression to the mean: The process first noted by Sir Francis Galton that "each peculiarity in man is shared by his kinsmen, but on the average to a less degree." Hence the tendency, for example, for tall parents to produce tall offspring but who, on the average, are shorter than their parents.

Research hypothesis: Synonym for *alternative hypothesis*.

Response variable: The variable of primary importance in psychological investigations, because the major objective is usually to study the effects of treatment and/or other explanatory variables on this variable and to provide suitable models for the relationship between it and the explanatory variables.

Robust statistics: Statistical procedures and tests that still work reasonably well even when the assumptions on which they are based are mildly (or perhaps moderately) violated. *Student's t test*, for example, is robust against departures from normality.

Rounding: The procedure used for reporting numerical information to fewer decimal places than used during analysis. The rule generally adopted is that excess digits are simply discarded if the first of them is smaller than five; otherwise the last retained digit is increased by one. Thus rounding 127.249341 to three decimal places gives 127.249.

S

Sample: A selected subset of a population chosen by some process, usually with the objective of investigating particular properties of the parent population.

Sample size: The number of individuals to be included in an investigation, usually chosen so that the study has a particular *power* of detecting an effect of a particular size. Software is available for calculating sample size for many types of study.

Sampling distribution: The *probability distribution* of a statistic. For example, the sampling distribution of the arithmetic mean of samples of size n taken from a *normal distribution* with mean μ and standard deviation σ is a normal distribution also with mean μ but with standard deviation $\sigma/\sqrt{n}$.

Sampling error: The difference between the sample result and the population characteristic being estimated. In practice, the sampling error can rarely be determined because the population characteristic is not usually known. With appropriate sampling procedures, however, it can be kept small and the investigator can determine its probable limits of magnitude. See also *standard error*.

Sampling variation: The variation shown by different samples of the same size from the same population.

Saturated model: A model that contains all *main effects* and all possible *interactions* between factors. Because such a model contains the same number of parameters as observations, it results in a perfect fit for a data set.

Scatter diagram: A two-dimensional plot of a sample of bivariate observations. The diagram is an important aid in assessing what type of relationship links the two variables.

SE: Abbreviation for *standard error*.

Semi-interquartile range: Half the difference between the upper and lower quartiles.

Sequential sums of squares: A term encountered primarily in regression analysis for the contributions of variables as they are added to the model in a particular sequence. Essentially the difference in the residual sum of squares before and after adding a variable.

Significance level: The level of probability at which it is agreed that the *null hypothesis* will be rejected. It is conventionally set at .05.

Significance test: A statistical procedure that when applied to a set of observations results in a *p value* relative to some hypothesis. Examples include the *Student's t test, the z test,* and *Wilcoxon's signed rank test*.

Singular matrix: A *square matrix* whose determinant is equal to zero; a matrix whose inverse is not defined.

Skewness: The lack of symmetry in a *probability distribution*. Usually quantified by the index, *s*, given by

$$s = \frac{\mu_3}{\mu_2^{3/2}},$$

where μ_2 and μ_3 are the second and third moments about the mean. The index takes the value zero for a symmetrical distribution. A distribution is said to have positive skewness when it has a long thin tail at the right, and to have negative skewness when it has a long thin tail to the left.

Split-half method: A procedure used primarily in psychology to estimate the reliability of a test. Two scores are obtained from the same test, either from alternative items, the so-called odd–even technique, or from parallel sections of items. The correlation of these scores or some transformation of them gives the required reliability. See also *Cronbach's alpha*.

Square contingency table: A contingency table with the same number of rows as columns.

Square matrix: A matrix with the same number of rows as columns. *Variance–covariance matrices* and *correlation matrices* are statistical examples.

Square root transformation: A transformation of the form $y = \sqrt{x}$ often used to make *random variables* suspected to have a *Poisson distribution* more suitable for techniques such as *analysis of variance* by making their variances independent of their means.

Standard deviation: The most commonly used measure of the spread of a set of observations. It is equal to the square root of the variance.

Standard error: The *standard deviation* of the *sampling distribution* of a statistic. For example, the standard error of the sample mean of *n* observations is $\sigma/\sqrt{n}$, where σ^2 is the variance of the original observations.

Standardization: A term used in a variety of ways in psychological research. The most common usage is in the context of transforming a variable by dividing by its *standard deviation* to give a new variable with a standard deviation of 1.

Standard normal variable: A variable having a *normal distribution* with mean zero and variance one.

Standard scores: Variable values transformed to zero mean and unit variance.

Statistic: A numerical characteristic of a sample, for example, the sample mean and sample variance. See also *parameter*.

Student's *t* distribution: The *probability distribution* of the ratio of a normal variable with mean zero and standard deviation one, to the square root of a chi-squared variable. In particular, the distribution of the variable

$$t = \frac{\bar{x} - \mu}{s/\sqrt{n}},$$

where $\bar{x}$ is the arithmetic mean of n observations from a *normal distribution* with mean μ and s is the sample standard deviation. The shape of the distribution varies with n, and as n gets larger it approaches a standard normal distribution.

Student's *t* tests: Significance tests for assessing hypotheses about population means. One version is used in situations in which it is required to test whether the mean of a population takes a particular value. This is generally known as a *single sample t test*. Another version is designed to test the equality of the means of two populations. When independent samples are available from each population, the procedure is often known as the *independent samples t test* and the *test statistic* is

$$t = \frac{\bar{x}_1 - \bar{x}_2}{s\sqrt{(1/n_1) + (1/n_2)}},$$

where $\bar{x}_1$ and $\bar{x}_2$ are the means of samples of size n_1 and n_2 taken from each population, and s^2 is an estimate of the assumed common variance given by

$$s^2 = \frac{(n_1 - 1)s_1^2 + (n_2 - 1)s_2^2}{n_1 + n_2 - 2}.$$

where s_1^2 and s_2^2 are the two sample variances.
If the *null hypothesis* of the equality of the two population means is true, then t has a *Student's t distribution* with $n_1 + n_2 - 2$ degrees of freedom, allowing *p values* to be calculated. In addition to homogeneity, the test assumes that each population has a *normal distribution* but is known to be relatively insensitive to departures from this assumption. See also *matched pairs t test*.

Symmetric matrix: A *square matrix* that is symmetrical about its leading diagonal; that is, a matrix with elements a_{ij} such that $a_{ij} = a_{ji}$. In statistics, *correlation matrices* and *covariance matrices* are of this form.

T

Test statistic: A statistic used to assess a particular hypothesis in relation to some population. The essential requirement of such a statistic is a known distribution when the *null hypothesis* is true.

Tolerance: A term used in stepwise regression for the proportion of the sum of squares about the mean of an explanatory variable not accounted for by other variables already included in the regression equation. Small values indicate possible multicollinearity problems.

Trace of a matrix: The sum of the elements on the main diagonal of a *square matrix*; usually denoted as tr(**A**). So, for example, if **A**$=(\begin{smallmatrix}3&2\\4&1\end{smallmatrix})$ then tr(**A**) $= 4$.

Transformation: A change in the scale of measurement for some variable(s). Examples are the *square root transformation* and *logarithmic transformation*.

Two-sided test: A test in which the *alternative hypothesis* is not directional, for example, that one population mean is either above or below the other. See also *one-sided test*.

Type I error: The error that results when the *null hypothesis* is falsely rejected.

Type II error: The error that results when the *null hypothesis* is falsely accepted.

U

Univariate data: Data involving a single measurement on each subject or patient.

U-shaped distribution: A *probability distribution* of *frequency distribution* shaped more or less like a letter U, though it is not necessarily symmetrical. Such a distribution has its greatest frequencies at the two extremes of the range of the variable.

V

Variance: In a population, the second moment about the mean. An *unbiased* estimator of the population value is provided by s^2 given by

$$s^2 = \frac{1}{n-1} \sum_{i=1}^{n} (x_i - \bar{x})^2,$$

where $x_1, x_2, \ldots, x_n$ are the n sample observations and $\bar{x}$ is the sample mean.

Variance–covariance matrix: Synonymous with *covariance matrix*.

Vector: A matrix having only one row or column.

W

Wilcoxon's signed rank test: A distribution free method for testing the difference between two populations using matched samples. The test is based on the absolute differences of the pairs of characters in the two samples, ranked according to size, with each rank being given the sign of the difference. The test statistic is the sum of the positive ranks.

Z

z **scores:**   Synonym for *standard scores*.

z **test:**   A test for assessing hypotheses about population means when their variances are known. For example, for testing that the means of two populations are equal, that is $H_0 : \mu_1 = \mu_2$, when the variance of each population is known to be σ^2, the *test statistic* is

$$z = \frac{\bar{x}_1 - \bar{x}_2}{\sigma \sqrt{\frac{1}{n_1} + \frac{1}{n_2}}},$$

where $\bar{x}_1$ and $\bar{x}_2$ are the means of samples of size n_1 and n_2 from the two populations. If H_0 is true, then z has a *standard normal distribution*. See also *Student's t tests*.

Appendix B

Answers to Selected Exercises

CHAPTER 1

1.2. One alternative explanation is the systematic bias that may be produced by always using the letter Q for Coke and the letter M for Pepsi. In fact, when the Coca Cola company conducted another study in which Coke was put into *both* glasses, one labeled M and the other labeled Q, the results showed that a majority of people chose the glass labeled M in preference to the glass labeled Q.

1.4. The quotations are from the following people: 1, Florence Nightingale; 2, Lloyd George; 3, Joseph Stalin; 4, W. H. Auden; 5, Mr. Justice Streatfield; 6, Logan Pearsall Smith.

CHAPTER 2

2.3. The graph in Figure 2.34 commits the cardinal sin of quoting data out of context; remember that graphics often lie by omission, leaving out data sufficient for comparisons. Here a few more data points for other years in the area would be helpful, as would similar data for other areas where stricter enforcement of speeding had not been introduced.

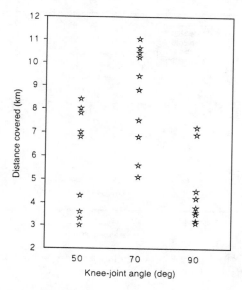

FIG. B.1.   Simple plot of data from ergocycle study.

CHAPTER 3

3.2. Here the simple plot of the data shown in Figure B1 indicates that the 90° group appears to contain two outliers, and that the observations in the 50° group seem to split into two relatively distinct classes. Such features of the data would, in general, have to be investigated further before any formal analysis was undertaken.

The ANOVA table for the data is as follows.

Source	SS	DF	MS	F	p
Between angles	90.56	2	45.28	11.52	.002
Within angles	106.16	27	3.93		

Because the groups have a clear ordering, the between group variation might be split into components, with one degree of freedom's representing variation that is due to linear and quadratic trends.

3.3. The term μ in the one-way ANOVA model is estimated by the grand mean of all the observations. The terms α_i, $i = 1, 2, \ldots, k$ are estimated by the deviation of the appropriate group mean from the grand mean.

3.5. The ANOVA table for the discharge anxiety scores is as follows.

Source	SS	DF	MS	F	p
Between methods	8.53	2	4.27	2.11	.14
Within methods	54.66	27	2.02		

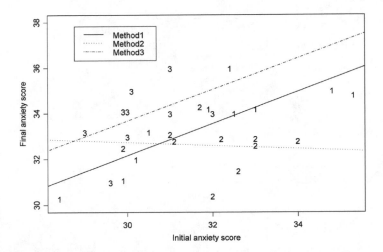

FIG. B.2. Plot of final versus initial anxiety scores for wisdom tooth extraction data.

Figure B2 shows a plot of final anxiety score against initial anxiety score, with the addition of the regression line of final on initial for each method. The line for method 2 appears to have a rather different slope from the other two lines. Consequently, an analysis of covariance may not be justified.

If, however, you ignore B2, then the analysis of covariance results are as follows.

Source	SS	DF	MS	F	p
Initial anxiety	6.82	1	6.82	4.87	.036
Between methods	19.98	2	9.99	7.13	.003
Within methods	36.40	26	1.40		

The analysis of variance for the difference, final score − initial score, is as follows.

Source	SS	DF	MS	F	p
Between methods	48.29	2	24.14	13.20	.0000
Within methods	49.40	27	1.83		

A suitable model that allows both age and initial anxiety to be covariates is

$$y_{ij} = \mu + \alpha_i + \beta_1(x_{ij} - \bar{x}) + \beta_2(z_{ij} - \bar{z}) + \epsilon_{ijk},$$

where x and z represent initial anxiety and age, respectively. The ANCOVA results from this model are as follows.

Source	SS	DF	MS	F	p
Initial anxiety	6.82	1	6.82	4.72	.039
Age	0.78	1	0.78	0.54	.468
Between methods	19.51	2	9.75	6.76	.004
Within methods	36.08	25	1.44		

CHAPTER 4

4.1. Try a logarithmic transformation.

4.2. The 95% confidence interval is given by

$$(\bar{x}_1 - \bar{x}_2) \pm t_{12}s\sqrt{(1/n_1) + (1/n_2)},$$

where $\bar{x}_1$ and $\bar{x}_2$ are the mean weight losses for novice and experienced slimmers both using a manual, s is the square root of the error mean square from the ANOVA of the data, t_{12} is the appropriate value of Student's t with 12 degrees of freedom, and n_1 and n_2 are the sample sizes in the two groups. Applying this formula to the numerical values gives

$$[-6.36 - (-1.55)] \pm 2.18\sqrt{3.10}\sqrt{1/4 + 1/4},$$

leading to the interval $[-7.52, -2.10]$.

4.3. The separate analyses of variance for the three drugs give the following results.

Drug X

Source	SS	DF	MS	F	p
Diet (D)	294.00	1	294.00	2.32	.14
Biofeed (B)	864.00	1	864.00	6.81	.02
D × B	384.00	1	384.00	3.03	.10
Error	2538.00	20	126.90		

Drug Y

Source	SS	DF	MS	F	p
Diet (D)	3037.50	1	3037.50	21.01	<.001
Biofeed(B)	181.50	1	181.50	1.26	.28
D × B	541.50	1	541.50	3.74	.07
Error	2892.00	20	144.60		

Drug Z

Source	SS	DF	MS	F	p
Diet (D)	2773.50	1	2773.50	13.97	.001
Biofeed(B)	1261.50	1	1261.50	6.36	.02
D × B	181.50	1	181.50	0.91	.35
Error	3970.00	20	198.50		

The separate analyses give nonsignificant results for the diet × biofeed interaction, although for drugs X and Y, the interaction terms approach significance at the 5% level. The three-way analysis given in the text demonstrates that the nature of this interaction is *different* for each drug.

4.11. The main effect parameters are estimated by the difference between a row (column) mean and the grand mean. The interaction parameter corresponding to a particular cell is estimated as (cell mean − row mean − column mean − grand mean).

4.12. An ANCOVA of the data gives the following results.

Source	SS	DF	MS	F	p
Pretreatment 1	8.65	1	8.65	0.66	.43
Pretreatment 2	35.69	1	35.69	2.72	.12
Between treatments	27.57	1	27.57	2.10	.17
Error	222.66	17	13.10		

CHAPTER 5

5.3. The ANOVA table for the data is

Source	SS	DF	MS
Between subjects	1399155.00	15	93277.00
Between electrodes	281575.43	4	70393.86
Error	1342723.37	60	22378.72

Thus using the nomenclature introduced in the text, we have

$$\hat{\sigma}_t^2 = \frac{93277.00 - 22378.72}{5} = 14179.66,$$

$$\hat{\sigma}_0^2 = \frac{70393.86 - 22378.72}{16} = 3000.95,$$

$$\hat{\sigma}_\epsilon^2 = 22378.72.$$

So the estimate of the intraclass correlation coefficient is

$$R = \frac{14179.66}{14179.66 + 3000.95 + 22378.72} = 0.358.$$

A complication with these data is the two extreme readings on subject 15. The reason given for such extreme values was that the subject had very hairy arms! Repeating the ANOVA after removing this subject gives the following results.

Source	SS	DF	MS
Between subjects	852509.55	14	60893.54
Between electrodes	120224.61	4	30056.15
Error	639109.39	56	11412.67

The intraclass correlation coefficient is now 0.439.

CHAPTER 6

6.1. Figure B3 shows plots of the data with the following models fitted: (a) log (vocabulary size) $= \beta_0 + \beta_1 \text{age}$; (b) vocabulary size $= \beta_0 + \beta_1 \text{age} + \beta_2 \times 2\text{age}^2$; and (c) vocabulary size $= \beta_0 + \beta_1 \text{age} + \beta_2 \times 2\text{age}^2 + \beta_3 \text{age}^3$.

It appears that the cubic terms model provides the best fit. Examine the residuals from each model to check.

6.4. Figure B4 shows plots of (a) p versus t; (b) log p versus t; and (c) p versus log t. One might imagine that a possible model here is $p = \exp(-\beta t)$, suggesting geometric loss of memory. Such a model should result in linearity in a plot of log p versus t, that is; plot (b) in Figure B4, but plainly this does not happen. A model that does result in linearity is to plot p against log t, although this is more difficult to explain.

6.5. Introduce two dummy variables x_1 and x_2 now defined as follows.

	Group		
	1	2	3
x_1	1	0	0
x_2	0	1	0

A multiple regression using these two variables gives the same ANOVA table as that in Display 6.9, but the regression coefficients are now the differences between a group mean and the mean of Group 3.

CHAPTER 7

7.1. Using the maximum score as the summary measure for the lecithin trial data and applying a t test gives these results:
$t = -2.68$; 45 degrees of freedom; $p = .0102$. A 95% confidence interval is $(-6.85, -0.98)$.

7.5. Fitting the specified model with now the Placebo group coded as -1 and the Lecithin group as 1 gives the following results.

Fixed effects

Parameters	Estimate	SE	t	p
Intercept	8.21	0.72	11.35	<.0001
Group	−2.37	0.72	−3.28	.002
Visit	0.58	0.12	4.73	<.0001
Group × Visit	1.301	0.12	10.68	<.0001

Random effects

$\hat{\sigma}_a = 4.54$, $\hat{\sigma}_b = 0.61$, $\hat{\sigma}_{ab} = -0.47$, $\hat{\sigma}_\epsilon = 1.76$

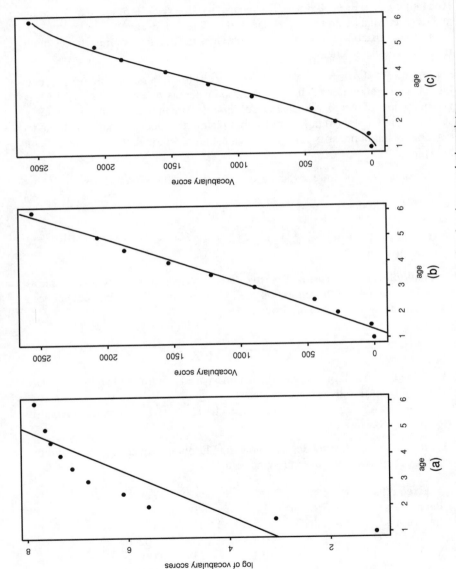

FIG. B.3. Residual from three models fitted to the vocabulary data.

364

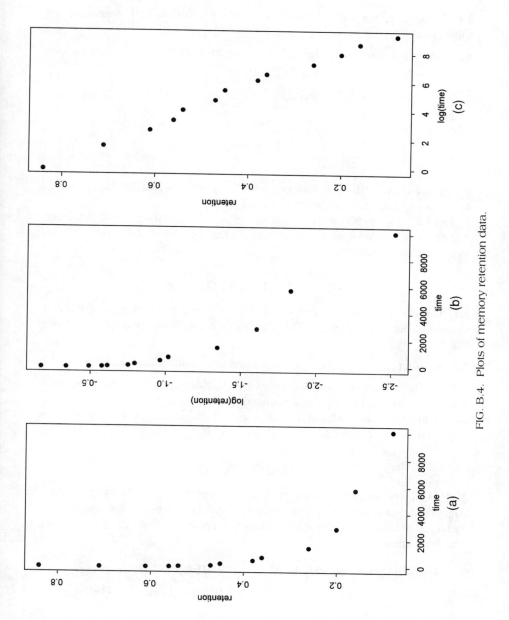

FIG. B.4. Plots of memory retention data.

CHAPTER 8

8.1. The signed rank statistic takes the value 40 with $n = 9$. The associated p value is 0.0391.

8.3. The results obtained by the author for various numbers of bootstrap samples are as follows.

N	95% Confidence Interval
200	1.039, 7.887
400	1.012, 9.415
800	1.012, 8.790
1000	1.091, 9.980

8.7. Pearson's correlation coefficient is -0.717, with $t = -2.91$, DF $= 8$, $p = .020$ for testing that the population correlation is zero.

Kendall's τ is -0.600, with $z = -2.415$, $p = .016$ for testing for independence.

Spearman's correlation coefficient is -0.806, with $z = -2.44$, $p = .015$ for testing for independence.

CHAPTER 9

9.1. The chi-squared statistic is 0.111, which with a single degree of freedom has an associated p value of 0.739. The required confidence interval is $(-0.042, 0.064)$.

9.3. An exact test for the data gives a p value of less than .000001.

9.8. Here because of the paired nature of the data, McNemar's test should be used. The test statistic is 4.08 with an associated p value of less than .05. The mothers' ratings appear to have changed, with a greater proportion of children rated as improved in year 2 than rated as not doing as well.

CHAPTER 10

10.1. An Investigation of a number of logistic regression models, including ones containing interactions between the three explanatory variables, shows that a model including each of the explanatory variables provides an adequate fit to the data. The fitted model is

$$\ln(\text{odds of being severely hurt}) = -0.9401 + 0.3367(\text{weight})$$
$$+ 1.030(\text{ejection}) + 1.639(\text{type}),$$

where weight is 0 for small and 1 for standard, ejection is 0 for not ejected and 1 for ejected, and type is 0 for collision and 1 for rollover. The 95% confidence intervals for the conditional odds ratios of each explanatory variable are weight, 1.18, 1.66; ejection, 2.31, 3.40, type, 4.38, 6.06. Thus, for example, the odds of being severely hurt in a rollover accident are between 4.38 and 6.06 those for a collision.

10.2. With four variables there are a host of models to deal with. A good way to begin is to compare the following three models: (a) main effects model, that is, Brand + Prev + Soft + Temp; (b) all first-order interactions, that is, model (A) above + Brand. Prev + Brand. Soft + Brand. Temp + Prev. Soft + Prev. Temp + Soft. Temp; (c) all second-order interactions, that is, model (B) above + Brand. Prev. Soft + Brand. Prev. Temp + Brand. Soft. Temp + Prev. Soft. Temp.

The likelihood ratio goodness-of-fit statistics for these models are as follows.

Model	LR Statistic	DF
A	42.93	18
B	9.85	9
C	0.74	2

Model A does not describe the data adequately but models B and C do. Clearly, a model more complex than A, but possibly simpler than B, is needed. Consequently, progress in searching for the best model might be made either by forward selection of interaction terms to add to A or by backward elimination of terms from B.

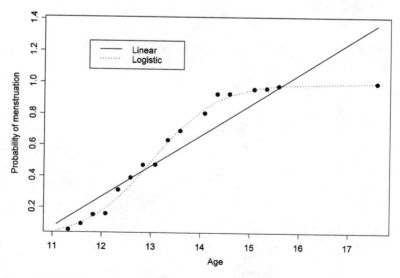

FIG. B.5. Linear and logistic models for probability of menstruation.

10.8. The plot of probability of menstruation with the fitted linear and logistic regressions is shown in Figure B5. The estimated parameters (etc.) for the two models are as follows.

Linear
Parameter	Estimate	SE	t
Intercept	−2.07	0.257	−8.07
Age	0.20	0.019	10.32

Logistic
Parameter	Estimate	SE	t
Intercept	−19.78	0.848	−23.32
Age	1.52	0.065	23.38

References

Agresti, A. (1996). *An introduction to categorical data analysis*. New York: Wiley.

Aitkin, M. (1978). The analysis of unbalanced cross-classification. *Journal of the Royal Statistical Society, Series A, 141*, 195–223.

Andersen, B. (1990). *Methodological errors in medical research*. Oxford: Blackwell Scientific.

Anderson, N. H. (1963). Comparisons of different populations: resistance to extinction and transfer. *Psychological Review, 70*, 162–179.

Bertin, J. (1981). *Semiology of graphics*. Wisconsin: University of Wisconsin Press.

Bickel, P. J., Hammel, E. A., and O'Connell, J. W. (1975). Sex bias in graduate admissions. Data from Berkeley. *Science, 187*, 398–404.

Boniface, D. R. (1995). *Experiment design and statistical methods for behavioural and social research*. London: Chapman and Hall.

Box, G. E. P. (1954). Some theorems on quadratic forms applied in the study of analysis of variance problems. II. Effects of inequality of variance and of correlations between errors in the two-way classification. *Annals of Mathematical Statistics, 25*, 484–498.

Brown, H., and Prescott, R. (1999). *Applied mixed models in medicine*. Chichester, UK: Wiley.

Carr, D. B. (1998). Multivariate Graphics. In P. Armitage and T. Colton (Eds.), *Encyclopedia of biostatistics*. Chichester, UK: Wiley.

Chambers, J. M., Cleveland, W. S., Kleiner, B., and Tukey, P. A. (1983). *Graphical methods for data analysis*. London: Chapman and Hall/CRC.

Chatterjee, S., and Price, B. (1991). *Regression analysis by example*. New York: Wiley.

Cleveland, W. S. (1985). *The elements of graphing data*. Summit, NJ: Hobart.

Cleveland, W. S. (1994). *Visualizing data*. Summit, NJ: Hobart.

Cochran, W. G. (1954). Some methods for strengthening the common chi-square tests. *Biometrics, 10,* 417–451.

Cohen, J. (1960). A coefficient of agreement for nominal scales. *Educational and Psychological Measurement, 20,* 37–46.

Collett, D. (1991). *Modelling binary data.* London: Chapman and Hall/CRC.

Colman A. M. (Ed.) (1994). *The companion encyclopedia of psychology.* London: Routledge.

Cook, R. D., and Weisberg, S. (1982). *Residuals and influence in regression.* London: Chapman and Hall/CRC.

Crutchfield, R. S. (1938). Efficient factorial design and analysis of variance illustrated in psychological experimentation. *Journal of Psychology, 5,* 339–346.

Crutchfield, R. S., and Tolman, E. C. (1940). Multiple variable design for experiments involving interaction of behaviour. *Psychological Review, 47,* 38–42.

Daly, F., Hand, D. J., Jones, M. C., Lunn, A. D., and McConway, K. J. (1995). *Elements of Statistics,* Wokingham: Addison-Wesley.

Davidson, M. L. (1972). Univariate versus multivariate tests in repeated measures experiments. *Psychological Bulletin, 77,* 446–452.

Davis, C. (1991). Semi-parametric and non-parametric methods for the analysis of repeated measurements with applications to clinical trials. *Statistics in Medicine, 10,* 1959–1980.

Diggle, P. J., Liang, K., and Zeger, S. L. (1994). *Analysis of longitudinal data.* Oxford: Oxford University Press.

Dizney, H., and Gromen, L. (1967). Predictive validity and differential achievement on three MLA comparative Foreign Language Tests. *Educational and Psychological Measurement, 27,* 1127–1130.

Donner, A., and Eliasziw, M. (1987). Sample size requirements for reliability studies. *Statistics in Medicine, 6,* 441–448.

Dunn, G. (1989). *Design and analysis of reliability studies: Statistical evaluation of measurement errors.* London: Arnold.

Efron, B., and Tibshirani, R. J. (1993). *An introduction to the bootstrap.* London: Chapman and Hall/CRC.

Eliasziw, M., and Donner, A. (1987). A cost-function approach to the design of reliability studies. *Statistics in Medicine, 6,* 647–656.

Everitt, B. S. (1968). Moments of the statistics kappa and weighted kappa. *British Journal of Mathematical and Statistical Psychology, 21,* 97–103.

Everitt, B. S. (1992). *The analysis of contingency tables.* (2nd edition) London: Chapman and Hall/CRC.

Everitt, B. S. (1998). *Cambridge dictionary of statistics,* Cambridge University Press, Cambridge.

Everitt, B. S., and Dunn, G. (2001). *Applied multivariate data analysis.* London: Arnold.

Everitt, B. S., and Pickles, A. (2000). *Statistical aspects of the design and analysis of clinical trials.* London: Imperial College Press.

Everitt, B. S., and Rabe-Hesketh, S. (2001). *The analysis of medical data using S-PLUS.* New York: Springer.

Everitt, B. S., and Wykes, T. (1999). *A dictionary of statistics for psychologists.* London: Arnold.

Fleiss, J. L. (1975). Measuring agreement between judges on the presence or absence of a trait. *Biometrics, 31,* 651–659.

Fleiss, J. L. (1986). *The design and analysis of clinical experiments.* New York: Wiley.

Fleiss, J. L., Cohen, J., and Everitt, B. S. (1969). Large sample standard errors of kappa and weighted kappa. *Psychological Bulletin, 72,* 323–327.

Fleiss, J. L., and Cuzick, J. (1979). The reliability of dichotomous judgements: unequal numbers of judgements per subject. *Applied Psychological Measurement, 3,* 537–542.

Fleiss, J. L., and Tanur, J. M. (1972). The analysis of covariance in psychopathology. In M. Hammer, K. Salzinger, and S. Sutton (Eds.), *Psychopathology.* New York: Wiley.

Friedman, M. (1937). The use of ranks to avoid the assumption of normality implicit in the analysis of variance. *Journal of the American Statistical Association, 32*, 675–701.

Frison, L., and Pocock, S. J. (1992). Repeated measures in clinical trials: analysis using means summary statistics and its implication for design. *Statistics in Medicine, 11*, 1685–1704.

Gardner, M. J., and Altman, D. G. (1986). Confidence intervals rather than P-values: estimation rather than hypothesis testing. *British Medical Journal, 292*, 746–750.

Gaskill, H. V., and Cox, G. M. (1937). Patterns in emotional reactions: I. Respiration; the use of analysis of variance and covariance in psychological data. *Journal of General Psychology, 16*, 21–38.

Goldberg, B. P. (1972). *The detection of psychiatric illness by questionnaire*. Oxford: Oxford University Press.

Goldstein, H. (1995). *Multilevel statistical models*. London: Arnold.

Good, P. (1994). *Permutation tests: A practical guide to resampling methods for testing hypotheses*. New York: Springer-Verlag.

Gorbein, J. A., Lazaro, G. G., and Little, R. J. A. (1992). Incomplete data in repeated measures analysis. *Statistical Methods in Medical Research, 1*, 275–295.

Greenacre, M. J. (1984). *Theory and application of correspondence analysis*. London: Academic Press.

Greenhouse, S. W., and Geisser, S. (1959). On the methods in the analysis of profile data. *Psychometrika, 24*, 95–112.

Haberman, S. J. (1974). *The analysis of frequency data*. Chicago: University of Chicago Press.

Hollander, M., and Wolfe, D. A. (1999). *Nonparametric statistical methods*. New York: Wiley.

Howell, D. C. (1992). *Statistical methods for psychology*. Belmont, CA: Duxbury Press.

Huynh, H., and Feldt, L. S. (1976). Estimates of the correction for degrees of freedom for sample data in randomised block and split-plot designs. *Journal of Educational Statistics, 1*, 69–82.

Johnson, V. E., and Albert, J. H. (1999). *Ordinal data modeling*. New York: Springer.

Kapor, M. (1981). *Efficiency on ergocycle in relation to knee-joint angle and drag*. Unpublished master's thesis, University of Delhi, Delhi.

Keselman, H. J., Keselman, J. C., and Lix, L. M. (1995). The analysis of repeated measurements: univariate tests, multivariate tests or both? *British Journal of Mathematical and Statistical Psychology, 48*, 319–338.

Krause, A., and Olson, M. (2000). *The basics of S and S-PLUS*. New York: Springer.

Krzanowski, W. J. (1991). *Principle of multivariate analysis*. Oxford: Oxford University Press.

Landis, J. R., and Koch, G. C. (1977). The measurement of observer agreement for categorical data. *Biometrics, 33*, 1089–1091.

Lindman, H. R. (1974). *Analysis of variance in complex experimental designs*. San Francisco, California: Freeman.

Lovie, A. D. (1979). The analysis of variance in experimental psychology: 1934–1945. *British Journal of Mathematical and Statistical Psychology, 32*, 151–178.

Lundberg, G. A. (1940). The measurement of socio-economic status. *American Sociological Review, 5*, 29–39.

Mann, H. B., and Whitney, D. R. (1947). On a test of whether one of two random variables is stochastically larger than the other. *Annals of Mathematical Statistics, 18*, 50–60.

Matthews, J. N. S. (1993). A refinement to the analysis of serial mesures using summary measures. *Statistics in Medicine, 12*, 27–37.

Maxwell, S. E., and Delaney, H. D. (1990). *Designing experiments and analysing data*. Belmont, CA: Wadsworth.

McCullagh, P., and Nelder, J. A. (1989). *Generalized linear models* (2nd ed.). London: Chapman and Hall.

McKay, R. J., and Campbell, N. A. (1982a). Variable selection techniques in discriminant analysis. I. Description. *British Journal of Mathematical and Statistical Psychology, 35*, 1–29.

McKay, R. J., and Cambell, N. A. (1982b). Variable selection techniques in discriminant analysis. II. Allocation. *British Journal of Mathematical and Statistical Psychology, 35*, 30–41.

McNemar, Q. (1962). *Statistical methods for research workers* (4th ed.). Edinburgh: Oliver and Boyd.

Morgan, G. A., and Griego, O. V. (1998). *Easy use and interpretation of SPSS for Windows*. Hillsdale, NJ: Erlbaum.

Nelder, J. A. (1977). A reformulation of linear models. *Journal of the Royal Statistical Society, Series A, 140*, 48–63.

Nelder, J. A., and Wedderburn, R. W. M. (1972). Generalized linear models. *Journal of the Royal Statistical Society, Series A, 155*, 370–384.

Nicholls, G. H., and Ling, D. (1982). Cued speech and the reception of spoken language. *Journal of Speech and Hearing Research, 25*, 262–269.

Novince, L. (1977). *The contribution of cognitive restructuring to the effectiveness of behavior rehearsal in modifying social inhibition in females*. Unpublished doctoral dissertation, University of Cincinnati.

Oakes, M. (1986). *Statistical inference: A commentary for the social and behavioural sciences*. Chichester: Wiley.

Quine, S. (1975). Achievement orientation of aboriginal and white Australian adolescents. Unpublished doctoral dissertation, Australian National University., Canberry.

Rawlings, J. O. (1988). *Applied regression analysis*. Belmont, CA: Wadsworth and Brooks.

Robertson, C. (1991). Computationally intensive statistics. In P. Lovie and A. D. Lovie (Eds.), *New developments in statistics for psychology and the social sciences* (pp. 49–80). London: BPS Books and Routledge.

Rosenthal, R., and Rosnow, R. T. (1985). *Contrast analysis*. Cambridge: Cambridge University Press.

Sauber, S. R. (1971). Approaches to precounseling and therapy training: an investigation of its potential inluence on process outcome. Unpublished doctoral dissertation, Florida State University, Tallahassce.

Schapiro, S. S., and Wilk, M. B. (1965). An analysis of variance test for normality (complete samples). *Biometrika, 52*, 591–611.

Schmid, C. F. (1954). *Handbook of graphic presentation*. New York: Ronald.

Schouten, H. J. A. (1985). Statistical measurement of interobserver agreement. Unpublished doctoral dissertation, Erasmus University, Rotterdam.

Schuman, H., and Kalton, G. (1985). Survey methods. In G. Lindzey and E. Aronson (Eds.), *Handbook of social psychology* (Vol. 1, p. 635). Reading, MA: Addison-Wesley.

Senn, S. J. (1994a). Repeated measures in clinical trials: analysis using mean summary statistics for design. *Statistics in Medicine, 13*, 197–198.

Senn, S. J. (1994b). Testing for baseline balance in clinical trials. *Statistics in Medicine, 13*, 1715–1726.

Senn, S. J. (1997). *Statistical issues in drug development*. Chichester: Wiley.

Singer, B. (1979). Distribution-free methods for non-parametric problems: a classified and selected bibliography. *British Journal of Mathematical and Statistical Psychology, 32*, 1–60.

Stevens, J. (1992). *Applied multivariate statistics for the social sciences*. Hillsdale, NJ: Erlbaum.

Theil, H. (1950). A rank-invariant method of linear and polynomial regession analysis. *I. Proc. Kon. Ned. Akad, v. Wetensch A, 53*, 386–392.

Tufte, E. R. (1983). *The visual display of quantitative information*. Cheshire, CT: Graphics.

Vetter, B. M. (1980). Working women scientists and engineers. *Science, 207*, 28–34.

Wainer, H. (1997). *Visual revelations*. New York: Springer-Verlag.

Westlund, K. B., and Kurland, L.T. (1953). Studies in multiple sclerosis in Winnipeg, Manitoba and New Orleans, Lousiana. *American Journal of Hygiene, 57*, 380–396.

Wilcoxon, F. (1945). Individual comparisons by ranking methods. *Biometrics Bulletin, 1*, 80–88.

Wilkinson, L. (1992). Graphical displays. *Statistical Methods in Medical Research, 1*, 3–25.

Williams, K. (1976). The failure of Pearson's goodness of fit statistic. *The Statistician, 25*, 49.

Index